RETRO ROCKETS

Experimental Rockets 1926-1941

Peter Alway

Robert H. Goddard's steering mechanism from his rocket L-15, launched May 19, 1937. Goddard pioneered rocket stabilization, using a gyroscope to control blast vanes (just behind the nozzle) and air vanes (one is being lifted away from the body). See page 39 details. (NASA photo 74-H-1228)

Retro Rockets: Experimental Rockets 1926-1941

ISBN 0-9627876-6-3

Published by:
Saturn Press
PO Box 3709
Ann Arbor, MI 48106-3709

The drawings in this book were produced on a Macintosh computer using MacDraw II and MacDraw Pro software.

The text was written with MacWrite II, and laid out with Ready Set Go.

Photographs were reproduced via photomechanical half-tones.

The body of the text was printed in 10-point Times font. Captions and drawings were printed with text in 9-point Helvetica.

This book is dedicated to the memory of

Clayton D. Alway

who taught me to appreciate books and gadgets.

Acknowledgements

The assistance of archivists, spacemodelers, and space collectors from around the world was essential to the writing of this book. I owe the following my gratitude:

Bob Alway, for insisting that the rockets of Goddard after 1926 were worth pursuing, and for directing me to the Goddard A-series rocket at the National Air and Space Museum's Garber Facility;

Stewart Bailey, for assistance in locating sources on German Rocketry;

C. James Cook, for providing food and lodging for my trip to the Goddard Library, for his help in reproducing the photograph of the Razumov-Shtern LRD-D-1, and for additional research at the Goddard Library;

Susan Dundas, for translating Russian material into English;

Riin Gill, for making it possible for me to write this book, and for diligently proofreading the text, and discovering and correcting hundreds of errors and inconsistencies;

Peter Gorin, for providing drawings of early Soviet rockets;

Yuri Hapon (Gapon), for providing drawings of early Soviet rockets;

David Howes, for providing data on the German A-3 and A-5;

Melissa A. N. Keiser of the National Air and Space Museum Archives, for her assistance in measuring and photographing the Goddard A-Series rocket at the Garber Facility;

Monica E. Milla, for suggesting the title, and in turn, the concept, of this book;

Alexander Mitiuriev, for providing a full-scale drawing of the GIRD-X;

Dorothy Mosakowski of the Clark University's Goddard Library Special Collections, for her help on locating descriptions, blueprints, and photographs of Goddard's rockets, and for her superhuman efforts to make my stay at Clark a pleasant one;

Michael Neufeld of the National Air and Space Museum, for documents and consultations on German rocketry;

Brian Nicklas of the National Air and Space Museum, for his considerable assistance above and beyond the call of duty in searching the Smithsonian Archives, locating numerous photographs, measuring original and replica Goddard rockets, and otherwise fulfilling the Smithsonian charter of disseminating human knowledge;

Robert Rau, of RMD Technology, Inc., for his assistance in printing text;

Lee Saegesser of the NASA History Office, for his assistance in locating photographs of early Soviet rockets;

André Siewa, for his heroic efforts in researching German rockets, including Johannes Winkler's HW-II, the A-3, and the A-5;

Bill Spadafora, for providing specifications of pipe used in ARS No. 4;

Glen Swanson, for contributing his motor home for a last-minute research trip to Washington;

Taras Tataryn, for providing data on Soviet rockets;

Dr. Andy Tomasch, for his assistance in interpreting drawings of the GIRD X;

Roger Wilfong, for providing an essential copy of Goddard's *Rocket Development.*

Robert H. Goddard's rocket L-10, launched December 18, 1936. This unguided rocket began a series of tests leading to several innovations in rocket control. See pages 36 and 39 for details. (courtesy of Clark University Goddard Library Archives)

Table of Contents

Introduction

Up to 1925, a rocket was a powder-burning projectile—a toy, really. While it had a history of competing with the cannon shell at various points in its history, it was not considered a means of transportation outside the world of science fiction. But in 1942, the rocket, in the form of the V-2, traveled through space and promised to become the ultimate weapon. In the years 1926 to 1941, tinkerers, professors, and governments transformed the rocket, in the process turning interplanetary travel from a fantasy to an engineering problem. This book is about the rockets that changed the world in a decade and a half.

The first purpose of this book is to open this early age of spaceflight to model makers. The drawings and photos will allow static modelers and model rocketeers to explore the era between Goddard's first liquid-propelled rocket and the V-2. The menagerie of rockets built over this period reflect naivete, poverty, unfocused creativity, and finally, solid engineering. This book concludes with plans for flying models of some of the rockets within. Certainly the peculiar forms of many of the rockets will add spice to any model rocket launch.

High power rocketry (HPR) enthusiasts may want to use the drawings of the full-size rockets themselves as plans. Many of the earliest experimental rockets were no larger than some popular HPR kits, and most flew no higher! HPR opens the door to a new hobby—replica rocketry, in which great moments in spaceflight history can be reenacted with full-scale replicas.

It is the aim of this book to bring the rockets themselves to the modeling public. The stories of the rocketeers of the 1930's are told to give the rockets a historical context. I would refer those who wish to know more of the space dreamers and doers of the era to Milton Lehman's *Robert H. Goddard: Pioneer of Space Research* (formerly published as *This High Man*), Willy Ley's *Rockets, Missiles, and Spaceflight,* Edward Pendray's *The Coming Age of Rocket Power,* James Oberg's *Red Star in Orbit,* Harry Wulforst's *The Rocket Makers,* Michael Neufeld's *The Rocket and the Reich,* and Frank Winter's *Prelude to the Space Age: The Rocket Societies 1926-1941.*

I have made an effort to provide drawings of every liquid-propelled vertical-ascent rocket flown before World War II, but there are naturally some gaps. In some cases it is not clear if a rocket ever flew. Soviet historical texts especially dwell on the innovations of native spaceflight pioneers often leaving readers to assume that rockets were built and flown. In other cases, there is simply no evidence on which to base a drawing (this is the case of the so-called "Curtain-Rod" rockets reportedly flown by a German amateur after the dissolution of the VfR).

A few notable solid-fueled rockets are included—chiefly because I found them irresistible and they have some historical significance—but I have made no effort to cover every notable solid-fueled rocket of the period, as they were not central to the story of spaceflight in the 1930's.

About the Drawings

There are few surviving engineering drawings of the rockets depicted in this book. I have been forced to rely on an odd collection of sources for dimensions and color information to reconstruct the appearances of these machines. Each group of rocketeers seems to have left their own set of challenges and contradictions for the space historian and modeler. These are outlined below as fair warning to the modeler and replica builder.

While Robert Goddard did not leave many engineering drawings of his rockets, he left extensive, explicit notes, and his wife Esther left hundreds of photographs of the dozens of liquid-fueled rockets flown by the professor. By 1931, Dr. Goddard's notes typically included detailed component-by-component descriptions, including dimensions of tanks and combustion chambers, and spacing between them. In most cases, it was possible to to work dimensions out along the entire length of a rocket. But inexplicably, the reported lengths of components and the spaces between them do not usually add up to the reported overall lengths. In these cases, the component lengths are drawn, and a note is made on the drawing alerting readers to the contradiction.

The drawing of Goddard's rocket of March 16, 1926 is based primarily on the replica displayed at the National Air and Space Museum (NASM), with some corrections based on photographs and Goddard's notes. While some of the original components of this rocket exist in an unflown rocket displayed at the National Air and Space Museum, this vehicle was not accessible for measurement.

Robert Goddard's notes occasionally included descriptions of color. From notes, photographs, and artifacts, one can gain an understanding of the professor's painting practices. Beginning in December of 1930 he painted one quadrant (the one facing the rear of the launch shelter, frustratingly out of the view of Esther Goddard's camera) Chinese red. By the launch of the first gyroscope-stabilized rocket in 1932, this included one side each of two fins. Sometimes certain components (oxygen tanks or nose caps) were left unpainted, but only rarely has specific evidence survived on this detail. Fortunately, the presence or absence of the red quadrant was explicitly mentioned in the notes on almost all of the early flights.

The 1935 A-Series rocket in storage at the National Air and Space Museum's Garber Facility has a full quadrant, including the liquid oxygen tank and nose cap, very neatly painted, and lacking better evidence, I have assumed this to be the paint pattern up to that time unless notes or photographs indicated otherwise. Certainly photographs indicate the neatness of the NASM artifact is atypical. Goddard's rockets seem to be hand-brushed, without benefit of masking.

Beginning in April of 1937, Goddard painted the other three quadrants of his rockets black to improve contrast in hazy skies. Goddard explicitly notes that the liquid oxygen tanks and nose caps of two rockets, L-14 and

Robert Goddard's test No. 72, the first to fly at Roswell, New Mexico, in the launch shelter prior to flight. This was the first liquid-propelled rocket to reach a significant altitude—2000 feet. (NASA photo 74-H-1194)

L-16 were not painted. Because these rockets shared components with rockets as early as Test L-10 in December of 1936, I have assumed that L-10 represents the beginning of unpainted oxygen tanks and nose cones. L-16 is the last rocket explicitly described as having one red and three black quadrants (L-17 was the same rocket re-flown).

The situation is cloudy for flights L-21 to L-30. A tail section has survived at least one of these flights—it is painted black on three quadrants, and unpainted on the fourth. Photographs of rocket components from this period show a mix of painted (presumably red) and unpainted contrasting quadrants. I am led to believe that the first of these rockets, L-21, sported a red quadrant, but that replacement parts used in later flights were only painted on the three black quadrants. The side of one of these rockets facing the rear of the launch tower must have displayed a wide stripe alternating red and silver. Unfortunately, all photographs of complete rockets of this period show the black side only.

Goddard's pump-fed rockets bore no red paint. The two that took flight were unpainted silver in one quadrant, and black in the other three. Goddard's last pump-fed rocket, which refused to leave the ground, was unpainted. This rocket is displayed at the National Air and Space Museum.

In some photographic reproductions, Goddard's nose cones look almost white. This is a photographic illusion—the spun aluminum nose caps were just a lighter shade of silver than the duralumin skin sections or the steel or nickel tanks.

The rockets of the American Rocket Society have been unevenly documented. Members produced widely published drawings of their rockets, but on close inspection the drawings don't match photographs of the rockets as flown. Drawings of ARS #2 were produced from memory, years after the flight, while drawings of ARS #4 represent an early configuration that did not fly. Fortunately, the paints applied to the fins of ARS #2 were recorded for posterity. Drawings in this book are based on photographs, with dimensions from drawings used only when they did not contradict the photos.

The drawings of German Rockets in this book are to be trusted the least. Only a few dimensions have survived the past 55 to 65 years, and many of them are suspect or blatantly wrong.

Very few dimensions of Johannes Winkler's rockets are known; the drawings are derived mainly from photographs. The color of the HW-1 is unknown, and only two B&W photographs are available; I have assumed unpainted metal. According to the notes of Tasillo Römisch, the HW-2 was silver and white, but it is not clear if these are based on anything but black and white photos.

The drawings of VfR rockets are mostly based on photographs, calibrated with a few dimensions listed in text sources. The 2-stick Repulsor, for instance, is based on a single photograph, and a description by Willy Ley. Ley provides no dimensions; as for color, he only mentions that the fins were unpainted aluminum. Other colors are educated guesses—the tanks look like steel in the photograph, sharp bends in the propellant lines suggest copper tubing, and the author's personal experience with small scientific plumbing fixtures is limited to brass. By comparison to Klaus Riedel in the photograph, it is evident that the tanks on this rocket were about 1 meter long. I have assumed that the tanks were exactly 1 meter long for the purposes of photographic interpretation.

The 1-Stick Repulsor was drawn from photographs, a passing description of color ("silver-painted") and a length and diameter printed in Michael Neufeld's *The Rocket and the Reich.* Neufeld was able to provide a German document with an engine length and nozzle diameter as well.

The 4-Stick Magdeburg Repulsor is based on photos, a few published dimensions, and a drawing of a tank cap found in the NASM archives. The rocket appears metallic, and is assumed to be silver.

Documentation on the A-2 is limited to a cutaway diagram of the rocket, a diameter, and a length. Color is anyone's guess. The fact that they were launched from a coastal island suggests they may have been painted as protection from corrosion.

The A-3 is based on a reasonable dimensioned drawing published in David Baker's *The Rocket.* This agrees within a few millimeters of another original Peenemünde drawing dating to 1942. While B&W photos show a dark color at Peenemünde and a metallic color at Kummersdorf during a static test, the indication of grey is based on nothing more than a photocopy from an anonymous hobby publication.

Most sources of diameter and length dimensions of the A-5 are questionable. In the end, I was forced to abandon English-language sources, no matter how scholarly, and use only photo-reproductions of original German drawings and photographs. Other sources were used only if they matched the drawings or measurements of photographs. The resulting drawing ignores the frequently reported enlargement of the fuselage diameter from the A-3; if this enlargement did take place, none of the reported lengths of the rocket in my sources can be correct. In a concession to this common wisdom, I have used the larger of two reported A-3 diameters for the A-5. Colors are from descriptions and photographs in Dieter Hölsken's *V-Missiles of the Third Reich.*

The Soviet drawings benefit from a long history of Soviet bloc model rocketry. Building and flying scale model rockets was a wholesome, state-sponsored youth activity in the Soviet Union, and dimensioned model and prototype drawings were widely distributed. As there was a special modeling contest category for pre-war Soviet rockets, they document early Soviet efforts remarkably well. It seems that engineering drawings of many early Soviet rockets have survived (apparently in the archives of the former Soviet Academy of Sciences), and even if they are not available in the West, they were available to the builders of museum replicas. These replicas seem to be the principal sources for many of the modelers' drawings. Occasional drawings have appeared in various conference proceedings published in the AAS History series. Generally the modelers' drawings are consistent with the less detailed drawings in the more scholarly sources. The case of the R-03 is less satisfactory, however, as the modeler's drawings do not agree well with photographs. The discrepency between the drawing and photo in this book is left for the modeler to puzzle out.

With a few exceptions, the drawings in this book are for the benefit of modelers reproducing the exterior configurations of the rockets, rather than their internal workings (just a few internal schematics are included to illuminate, but not as building guides). Hobbyists who wish to fly models and replicas of these rockets will be better served by the technology of model and high power rocketry than finicky and dangerous liquid propellant engines. History-minded engineers and museum replicators will find that tracking down the drawing sources and bibliographic references listed in this book is a trifling task next to building a complete, functional liquid-powered rocket.

Symbols, Units, and Abbreviations

Abbreviations:

Approx [10 (Approx)]—Approximately
ft [100 ft]—feet
in [6 in]—inches
kg [45 kg]—kilograms
lb [120 lb]—pounds
lb-sec [14,000 lb-s]—pound-seconds
m [1.6 m]—meters
mi [10 mi]—miles
mm [120 mm]—millimeters
N [1,200 N]—Newtons
N-sec [15,000 N-sec]—Newton-seconds
RAD [3.5 RAD]—Radius
sec [90 sec]—seconds
STA [STA 1237.3]—Station number.

Symbols:

~ [~10]—Approximately
Ø [12.0 Ø]—Diameter
' [6']—Feet
" [72"]—Inches

Metric and English Units

All vehicles were drawn in the units used by the originating country. To avoid conversion errors, modelers are advised to build in the system of the original rocket.

The inch is precisely defined in terms of the millimeter:

1" = 25.4 mm
1' = 304.8 mm
1 mm = ~.03937"
1 m = ~39.37"

STA (Station) Numbers

Distance from reference point (more precisely, reference plane), usually the tip of the nose or the base of a rocket. To use station numbers in model construction, locate the STA numbers from the front and rear of the part you wish to model. The difference between the numbers is the length of the part in the units specified below the drawing title.

NAR Designations

I have included thrust and weight data for some vehicles. In an effort to give modelers an intuitive sense of the power of these rockets, I have included NAR designations for rocket performance. In the NAR system, impulse doubles with each letter in the alphabet, beginning with the letter A at 2.5 Newton-Seconds (0.56 lb-sec). Since total impulse is more meaningful to rocket performance than thrust, this logarithmic system is ideal for describing many rocket vehicles. As in the NAR system, a rocket with an impulse lying between two letter values is assigned the higher letter value.

The Rockets of Robert H. Goddard

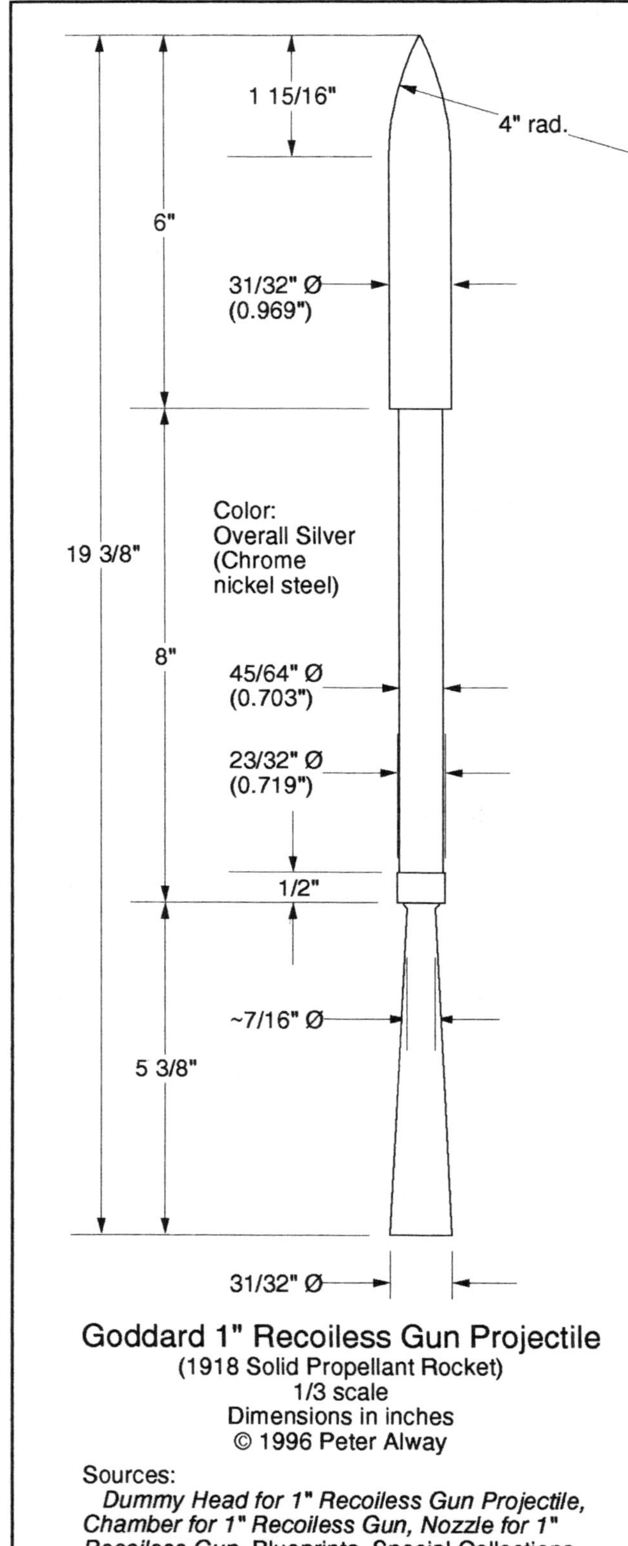

Goddard 1" Recoiless Gun Projectile
(1918 Solid Propellant Rocket)
1/3 scale
Dimensions in inches
© 1996 Peter Alway

Sources:
Dummy Head for 1" Recoiless Gun Projectile, Chamber for 1" Recoiless Gun, Nozzle for 1" Recoiless Gun, Blueprints, Special Collections, Goddard Library, 1918.
Approximate measurements of "Smokeless Powder Rocket" on display at National Air and Space Museum.

On October 19, 1899, a seventeen year old devotee of science fiction climbed a cherry tree in his family's orchard. As he set to trim its branches, he was overcome by a powerful fantasy: a great whirligig of a machine that would capture centrifugal force to one day transport him to Mars. This dream would possess Robert H. Goddard for the rest of his life.

As he progressed through his education in physics, Goddard realized that clever spinning perpetual motion machines would get him nowhere. By the time he started teaching at Clark University, in Massachusetts, Goddard recognized that rockets were the only way into space. He resolved to apply physics to the problem of the space rocket.

Experiments with Solid-Propellant Rockets

Goddard explored the mathematics of rocket propulsion, and realized that two factors drove a rocket's effectiveness. The first was the speed at which exhaust was ejected from the nozzle (divided by g, gravitational acceleration, this gives specific impulse, now favored as a definition of rocket efficiency). The second was the fraction of launch mass that was usable propellant.

Goddard tested various powder (solid propellant) rockets, and found them to convert no more than 2% of their energy into usable thrust. The most efficient, the Coston ship rocket, had a specific impulse of about 32 seconds (by contrast, the cheapest model rocket engines today have a specific impulse around 80 sec, and the Space Shuttle main engine has a specific impulse of 445 sec). The thermodynamics of the steam engine had been a hot topic in late 19th century physics. Goddard applied the steam turbine's de Laval nozzle (an expanding cone that used the expansion of exhaust gasses to do work) to the rocket. The result was not only the most efficient rocket to date, but the most efficient heat engine to date, converting over half the heat produced by burning into exhaust velocity. The resulting specific impulses approached 250 sec.

Goddard also demonstrated that rockets were not only capable of propulsion in the vacuum of space, but that they even performed better in a vacuum. In one demonstration, the professor loaded a 22-caliber revolver on a turntable mounted in an evacuated bell jar. He showed the recoil was equally effective in a vacuum and in air. In another test, he attached a rocket to the end of a long, looping, evacuated tube. With this apparatus he determined that his rockets could produce as much as 20% more thrust firing their exhaust into a vacuum.

Goddard put together a report on his method of reaching extreme altitudes, and submitted it to the Smithsonian Institution. In 1917, the Smithsonian began a long period of support for the rocket scientist with a $5000 grant.

With America's entry into the First World War, Goddard submitted proposals to the military. Early in 1918, he left for the Mount Wilson Observatory in California to

Goddard's 1" recoiless gun projectile of 1918, a small solid propellant rocket. The tag at the rear of the cylindrical section is not part of the rocket. (photo by author)

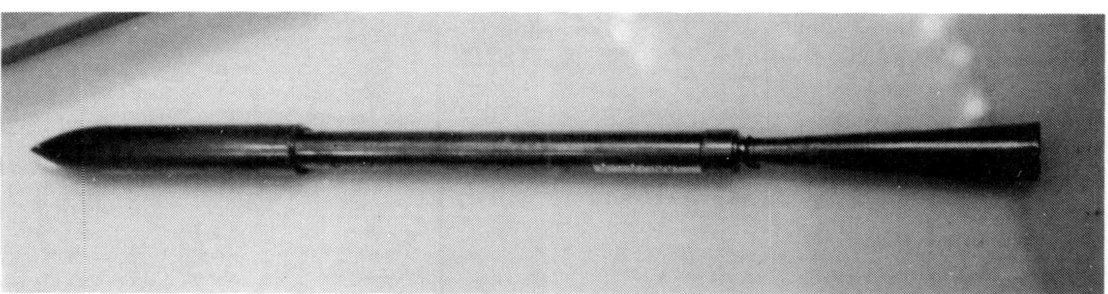

develop rocket projectiles for the Army. He followed two approaches: simple projectiles powered by single propellant grains, and ungainly multiple-charge contraptions that functioned by burning a succession of propellant chunks automatically fed into the combustion chamber in flight. Clarence Hickman, a graduate student, acted as Goddard's assistant. Hickman managed to put the multiple charge rocket into working order by altering it to fire charges into the chamber through the open nozzle below, rather than through an easily damaged blast door above. Eventually the multiple-charge rocket could blast itself upward a few hundred feet.

Goddard needed no help designing the single grain rockets. The smallest of these rockets was the 1" recoiless gun projectile, just under an inch (25 mm) in diameter (to clear the inside of a 1" launch tube) and under 2 feet (0.5 m) long. Internally, the rocket was designed around a single 29/62" (11.5 mm) rod of nitroglycerin, the largest solid grain available. The rocket was machined from nickel steel in three parts: a nose cone, combustion chamber, and de Laval nozzle. Goddard built variations on each component: longer and shorter combustion chambers, smooth and rifled (with spiral grooves) nozzles, and lighter and heavier nose cones. Launched by a 5 1/2-foot (1.7 m) tube, held on the shoulder or at the hip, one of these rockets flew a distance of 722 1/2 yards (661 m).

On November 6, 1918 Goddard and Hickman demonstrated their rockets for high Army officers at Aberdeen Proving Ground in Maryland. An assortment of 1", 1.75", and 3" (25 mm, 45 mm, and 76 mm) rocket tests were on the agenda. Hickman demonstrated the lack of recoil by launching projectiles from a tube balanced on a pair of music stands. One demonstration was of a "1" recoiless hand gun" with a projectile with a "heavy head." This particular version was to show the rocket in flight; presumably the extra weight would slow the rocket enough to make the rocket visible in the air. A rocket of this type survives, and is displayed at the National Air and Space Museum. The drawing opposite depicts this type.

Army officials were impressed by the professor's rockets, and showed a sincere interest in ordering them in quantity. But on November 7, 1918, the Germans surrendered, and four days later, World War I was formally over. Army interest evaporated immediately. The recoiless gun projectile would have to wait for the next war, when Hickman would turn it into the Bazooka.

In 1919, Goddard updated his proposal to the Smithsonian, and the Institution published it as "A Method of Reaching Extreme Altitudes," in 1920. Goddard described his experiments with high-efficiency solid-propellant rockets, including the ungainly multiple-charge rocket. Then with mathematical rigor, he extrapolated the results to demonstrate the possibility of high-altitude flight. Goddard concluded the report with a set of calculations proving the possibility of sending a rocket to the Moon. He suggested that such an accomplishment might be verified by launching a 13.82-lb (6.3 kg) load of flash powder (enough, he calculated, to be "strikingly visible") to the dark part of the new Moon. Not one for rounding off numbers, he calculated a launch weight of 33,278 lbs (15,000 kg)—just a bit smaller than the solid-fueled Scout that would launch 150 lbs (70 kg) into Earth orbit 40 years later.

In time, radio and television would render such a pyrotechnic display redundant, but the idea resurfaced in the Soviet Union during planning for the Luna probes in the late

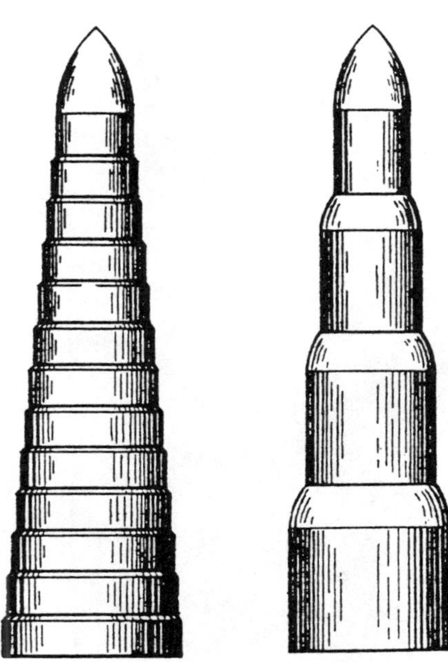

Goddard's Moon rocket: In discussing the potential of rockets for reaching extreme altitudes, Goddard considered the optimum number of stages for such a device. This figure from *A Method of Reaching Extreme Altitudes* shows two possible configurations for a multiple-stage Moon Rocket. These schematics were meant to illustrate the relative merits of many stages vs. few stages—resemblance to the Soviet N-1 and American Saturn V, respectively, is superficial. Goddard did not specify dimensions for these rockets, but based on their estimated weight, they might have been as small as 6 feet (2 m) in diameter and 24 feet (7 m) tall.

1950's. Recent reports indicate that the USSR considered detonating an atomic bomb on the Moon to insure the world would take note of the accomplishment.

Whether proposal or mere thought experiment, Goddard's lunar scheme transformed the author from an obscure physics professor to a celebrity mad scientist (Buck Rogers's Dr. Huer was based on Dr. Goddard). A January 13, 1920 *New York Times* editorial chided Goddard for demonstrating that a rocket worked in a vacuum.

> "As a method of sending a missile to the higher, and even highest, part of the earth's atmospheric envelope, Professor Goddard's multiple-charge rocket is a practicable, and therefore promising device...It is when one considers the multiple-charge rocket as a traveler to the moon that one begins to doubt...for after the rocket quits our air and and really starts on its longer journey, its flight would be neither accelerated nor maintained by the explosion of the charges it then might have left. To claim that it would be is to deny a fundamental law of dynamics, and only Dr. Einstein and his chosen dozen, so few and fit, are licensed to do that."
>
> "That Professor Goddard, with his 'chair' in Clark College and the countenancing of the Smithsonian Institution, does not know the relation of action to reaction, and of the need to have something better than a vacuum against which to react—to say that would be absurd. Of course he only seems to lack the knowledge ladled out daily in high schools..."

The *Times* retracted the editorial in 1969.

The First Liquid Propellant Rocket

In 1921, Goddard gave up on his multiple cartridge solid propellant system. The failure of this awkward device led him to a conclusion that Tsiolkovsky had come to in Russia at the turn of the century. Space travel would require liquid propellant rockets.

For five years, Goddard pleaded for money, tested gasoline/liquid oxygen rocket engines, re-designed, re-tested, and pleaded for more money. Goddard had managed to extract a grant for small-scale work from the Smithsonian Institution, but work was slow. Liquid-fueled rockets would require special pumps, and the professor's small piston pumps were just not working out. Late in 1925, he abandoned pumps and tried a simple gas pressure feed system. Pumps could wait—an actual flight demonstration could bring the financial support for larger vehicles to follow.

Goddard spent the winter tweaking the flow rates of gasoline and oxygen into his 2" (50 mm) diameter combustion chamber. When he found that combustion took place in the nozzle, he re-designed the injection system with multiple gasoline orifices, insuring complete burning within the combustion chamber. Finally, on January 20, 1926, a liquid-fueled rocket produced more thrust than its weight.

Goddard had built gleaming aluminum shrouds for his rocket (the nose of the forward shroud even held a silk parachute), but he abandoned them to save weight. The rocket he took out to his Aunt Effie Ward's farm on March 6 was an open collection of tanks and tubes. At the top was a combustion chamber equipped with a de Laval nozzle. Liquid oxygen (LOX) and gasoline flowed up from tanks at the base of the rocket through two asbestos-wrapped steel tubes. An asbestos-wrapped cone shielded the LOX tank from the engine's flame.

Goddard and his university machinist Henry Sachs ignited the rocket, but in four seconds, the bottom of the combustion chamber burned through. The nozzle shot off the back of the chamber, and the recoil jammed the rocket into the wire guides at the top of the tower, bending the propellant

Left: Esther Goddard took this photo of the first liquid-propelled rocket on March 6, 1926, before its first launch attempt. The rocket is rotated 180° from the proper launch orientation in this photo, and the launcher would be modified before the first flight. In spite of the misleading features, the photo has become famous for æsthetic reasons. (NASA photo 74-H-1065)

Goddard's Rocket of March 16, 1926

1/20 scale
Dimensions in inches

Sources:
NASA photo 74-H 1065.
Measurements of replica on display at National Air and Space Museum, Washington, DC.
Development of Liquid Propellant Rocket, 1921-1929, Robert H. Goddard, typed transcripts of notes, Special Collections, Goddard Library, section III,

Bare Metal (silver)

Polished metal (silver)

Asbestos (light Grey)

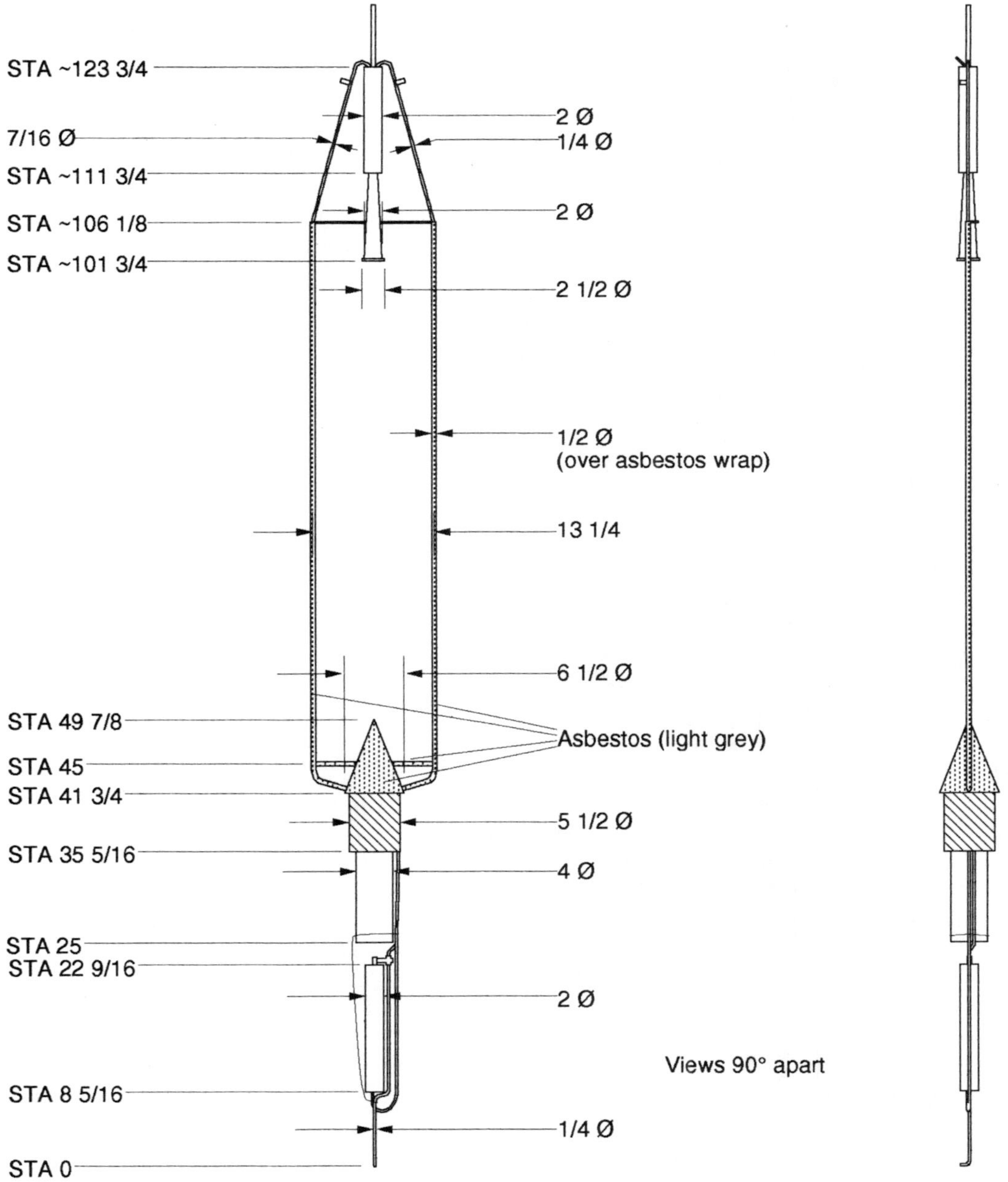

STA ~123 3/4
7/16 Ø
STA ~111 3/4
STA ~106 1/8
STA ~101 3/4
2 Ø
1/4 Ø
2 Ø
2 1/2 Ø
1/2 Ø
(over asbestos wrap)
13 1/4
6 1/2 Ø
STA 49 7/8
STA 45
STA 41 3/4
Asbestos (light grey)
5 1/2 Ø
STA 35 5/16
4 Ø
STA 25
STA 22 9/16
2 Ø
STA 8 5/16
1/4 Ø
STA 0

Views 90° apart

Launch Frame for Goddard's Rocket of March 16, 1926,

1/20 scale
Dimensions in inches

Sources:
Measurements of replica on display at National Air and Space Museum, Washington, DC.
Development of Liquid Propellant Rocket, 1921-1929, Robert H. Goddard, typed transcripts of notes, Special Collections, Goddard Library, section III,

Launch frame painted black.

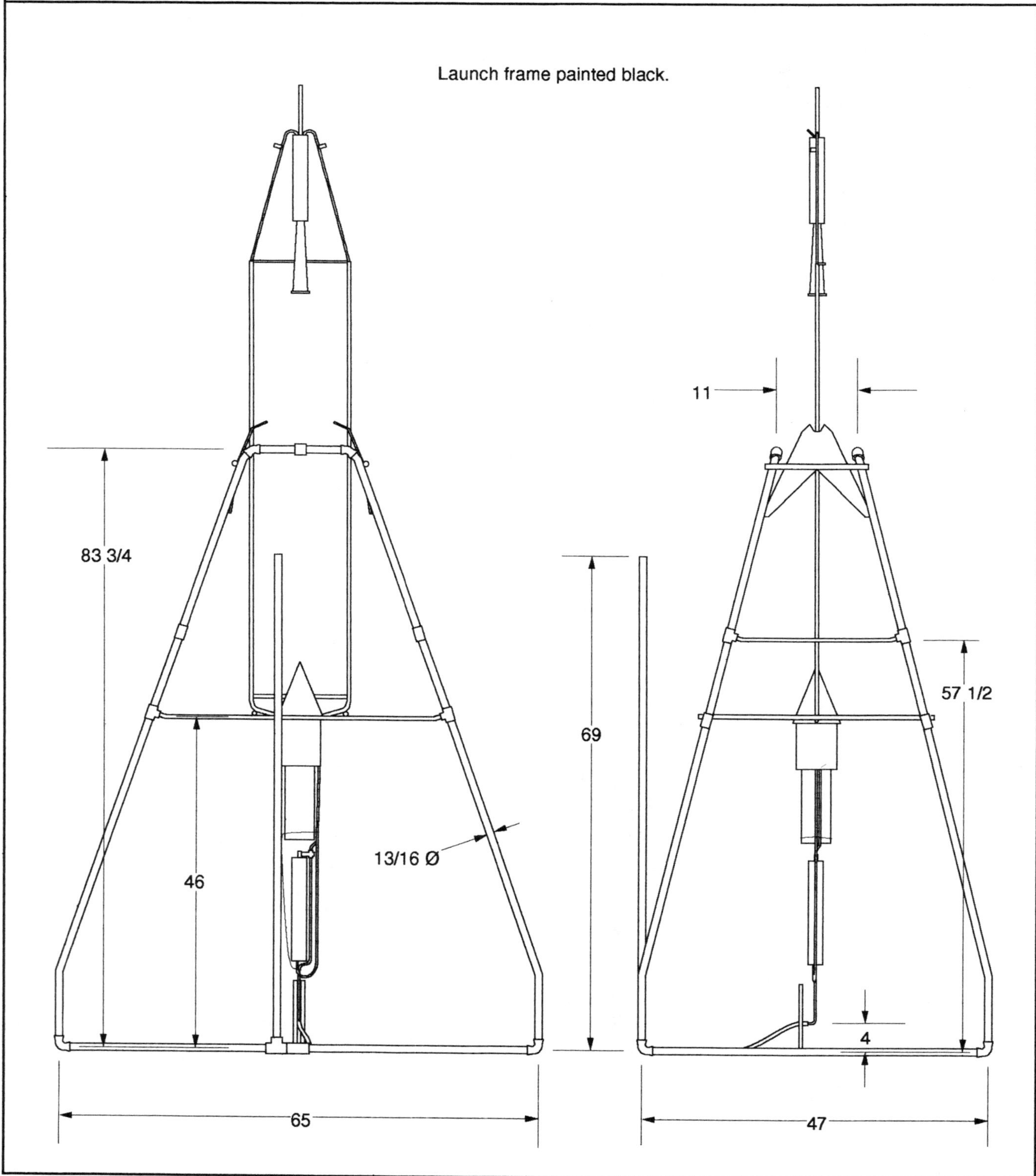

lines.

Goddard and Sachs returned to the shop, building a new nozzle and replacing the launcher guides with hinged sheets of galvanized iron that would not bind the rocket.

On March 16, Goddard and Sachs lugged the repaired rocket, along with the re-designed launch frame, to Effie Ward's farm. After they set up the rocket next to Aunt Effie's cabbage patch, Esther Goddard arrived with Assistant Professor Percy Roope, along with still and movie cameras.

After Esther photographed her husband beside his creation, Goddard and Sachs loaded gasoline and liquid oxygen into the rocket. Sachs lit a torch and held it to a pyrotechnic igniter protruding from the top of the combustion chamber. Once smoke poured from the chamber, he turned the pressure valve on the oxygen tank with the aid of a long stick. He then lit an alcohol burner—a bit of cotton in a cupped scrap of metal—under the oxygen tank. Goddard turned a valve to allow pressurized oxygen from a ground supply to enter the rocket, supplementing the oxygen vapor pressure with a 6 atmosphere jump start. The sudden increase in tank pressure forced liquid oxygen up the oxidizer line to the combustion chamber. A line connecting the two propellant tanks carried pressurized oxygen gas down to the gasoline tank, forcing its contents into the combustion chamber along the fuel line.

A small smokeless flame issued from the nozzle, but the rocket remained motionless. The Goddards watched as two thirds of the rocket nozzle burned away. Esther's movie camera ran down. Then, as Robert put it, the rocket decided "I think I'll get the hell out of here." It reached an altitude of 41 feet (12.3 m), and crashed into the snow 184 feet (55 m) away. Modern rocketry was born.

It took Robert Goddard just over two weeks to re-assemble the rocket for another flight. He reinforced the open frame of the rocket with two additional vertical segments of tubing and cross-braces, giving the machine a sturdy square cross-section. This rocket flew on April 3. In a final modification, he compressed the various components into a small rocket with the combustion chamber at the rear. Goddard tested the rocket on May 4, but it never flew (this rocket, now on display at the National Air and Space Museum, contains all that is left of the first liquid fueled rocket of March 16, 1926). He concluded the rocket, with its 2"-diameter (50 mm) combustion chamber, was too small for effective tinkering. He turned his attention to creating a larger rocket. But the result of a major scale-up of his previous rockets was a monster that could not lift its own weight. It was time to scale down again.

Above: The first liquid-propelled rocket to take flight, shown here on March 16, 1926. The guides at the top of the launcher differ slightly from those in a more famous photo taken a few days earlier. (courtesy of Clark University Goddard Library Archives)

Rocket of March 16, 1926 specifications

Loaded weight	10.45 lb (4.75 kg)
Empty weight	6 lb (2.7 kg)
Thrust	~18 lb (80 N)
Flight duration	2.5 sec
Total impulse	~45 lb-sec (200 N-sec)
NAR Designation	H80
Length	~10 ft (3 m)
Chamber Diameter	2 in (50 mm)

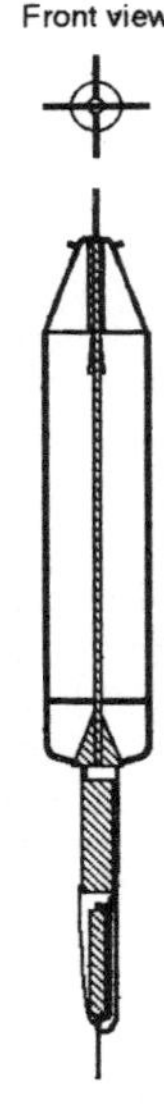

Right: The first liquid-fuelled rocket reconfigured for its second flight, on April 3, 1926, with dummy reinforcing tubes. (photo courtesy of Clark University Goddard Library Archives, 1/50 scale sketch by author)

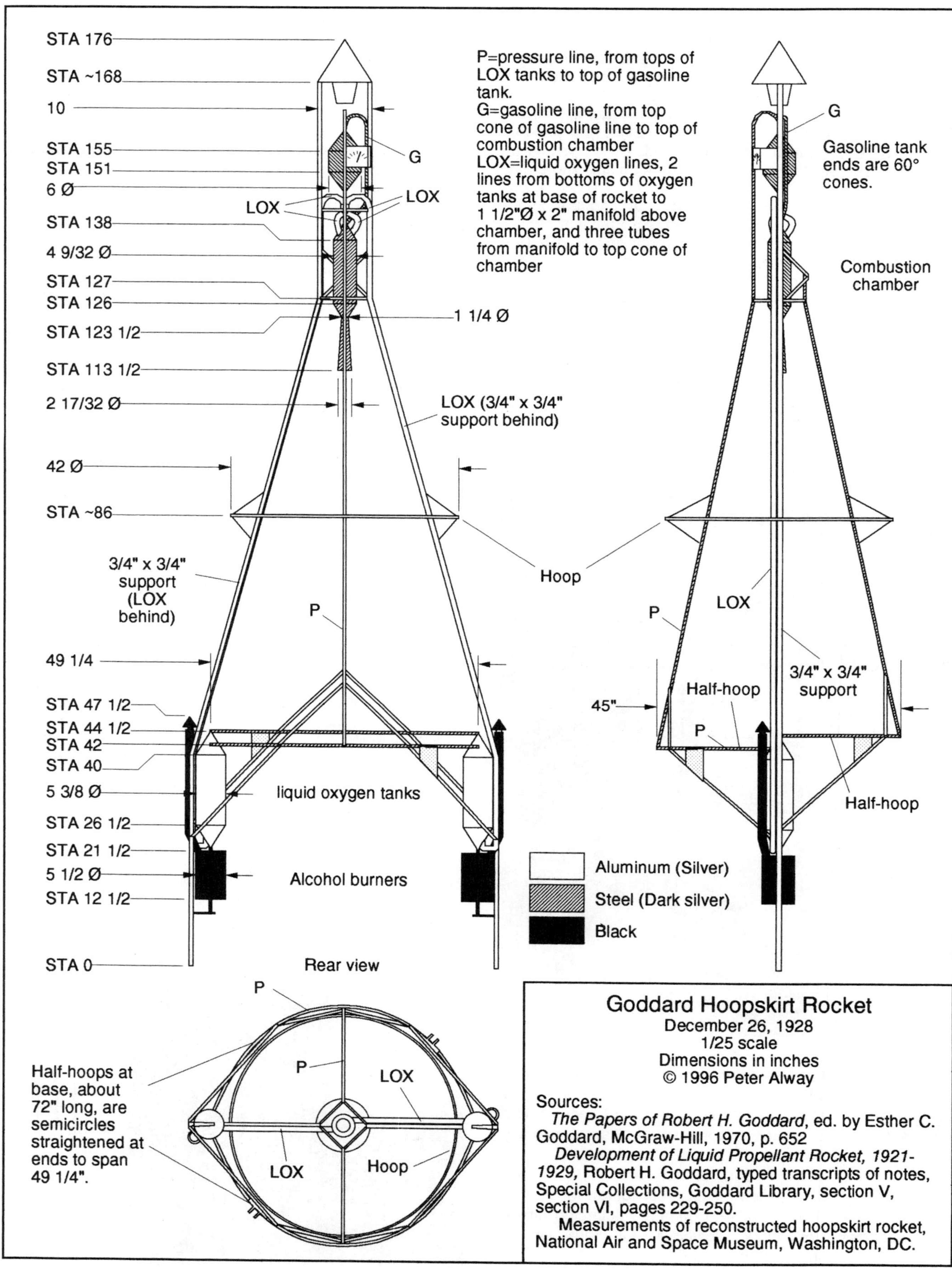

Goddard Hoopskirt Rocket

December 26, 1928
1/25 scale
Dimensions in inches

Sources:

The Papers of Robert H. Goddard, ed. by Esther C. Goddard, McGraw-Hill, 1970, p. 652

Development of Liquid Propellant Rocket, 1921-1929, Robert H. Goddard, typed transcripts of notes, Special Collections, Goddard Library, section V, section VI, pages 229-250.

Measurements of reconstructed hoopskirt rocket, National Air and Space Museum, Washington, DC.

The Hoopskirt Rocket

Goddard's next project would be his last tractor (nose-driven) rocket, which he dryly described as a "medium-sized rocket with combustion chamber above liquid oxygen tanks." It is better known as the "Hoopskirt" rocket. At the base of two long square tubes were a pair of liquid oxygen tanks. Below them were a pair of alcohol burners to heat the tanks and provide gas pressure to force the liquid up pipes to the combustion chamber. The gasoline tank above the combustion chamber was also pressurized by the oxygen vapor. The rocket took its name from a hoop halfway up the rocket and two half-hoops near the base that held the rocket components symmetrically out of the way of the rocket exhaust. By this time, Goddard understood that there was no stability advantage to placing the combustion chamber at the front of the rocket beyond the fact that it was the heaviest component of the rocket.

On July 18, 1929, Goddard and his assistants (now including graduate student Laurence Mansur and machinist Albert Kisk—Esther Goddard's brother) loaded the rocket into its launcher and ignited the engine. The rocket jammed in the tower. On September 29, Goddard and his assistants again loaded the rocket into its launcher and ignited the engine. The rocket jammed in the tower. On October 10, the rocket again jammed in the tower. Finally, on October 20, the rocket came to life, spewed flame, and began to rise under its own lift, and jammed in the tower again.

On December 26, 1928, the 28 1/2 lb (13 kg) rocket flew free of its launcher, promptly turned toward the observation shelter, flew over the Goddards' heads, and crashed near Mr. Mansur, who manned the theodolite. The rocket exploded on impact, setting a grass fire. While the path of the rocket made altitude measurement impossible, Mansur's maximum elevation measurement of 11.5° from 200 feet (60 m) indicates an altitude of no more than 41 feet (12 m), and probably closer to 20 feet (6 m).

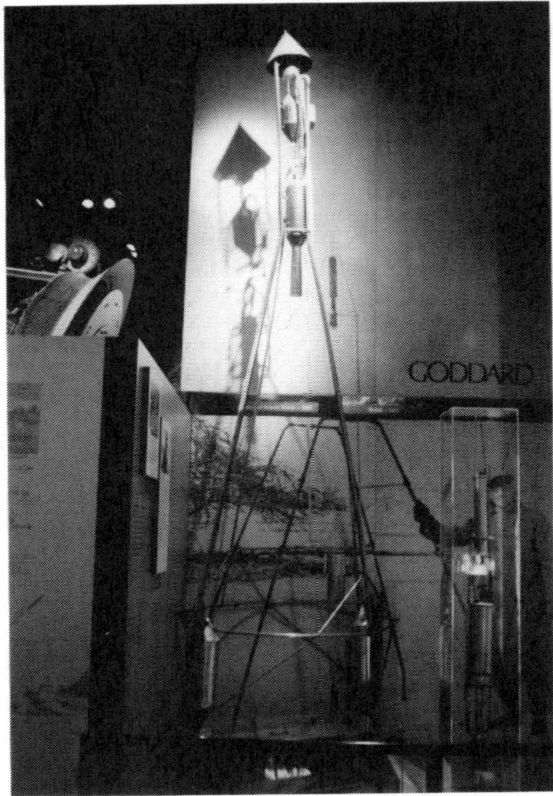

The original Hoopskirt rocket at the National Air and Space Museum. The upper hoop is missing from the rocket, reconstructed from wreckage. (photo by author)

The Rocket of July 17, 1929

Goddard's fourth rocket took flight on July 17, 1929. This machine featured an on-board camera to record data in flight. It was also Goddard's first tail-driven rocket, stabilized by four sheet metal fins. The roar, flaming flight, and crash of the rocket attracted the attention of one of Aunt Effie's neighbors, who reported a plane crash to the police. Goddard and company had cleared the wreckage, and were looking for one last part as a caravan of police, fire, and ambulance vehicles arrived. Over the professor's protests, the State Fire Marshal ended the tests. But not all the attention this test gathered was bad. Charles Lindbergh, fresh from his famous transatlantic flight of 1927, had been thinking about the future of aviation, and saw it as jet or rocket propelled. When he spotted an article, distorted as it was, on the Massachusetts professor's experiment, Lindbergh had to investigate.

Lindbergh had tried to interest companies in developing rocket propulsion for aircraft, but there was neither knowledge nor interest in rockets within the American aviation industry. Lindbergh met Goddard, and seeing that he was no crackpot, set about aiding the inventor in the best way he could. As a national hero, Lindbergh had the ear of men of great wealth. He asked Goddard the sort of support he would need to attain extreme altitudes. Robert calculated a budget of $25,000 per year for four years—an astronomical sum for the time.

It was not until June of 1930 that Lindbergh persuaded the millionaire and benefactor of aviation, Daniel Guggenheim, to part with $50,000 for a two-year program of rocket development. Goddard moved out to Roswell, New Mexico to continue his work.

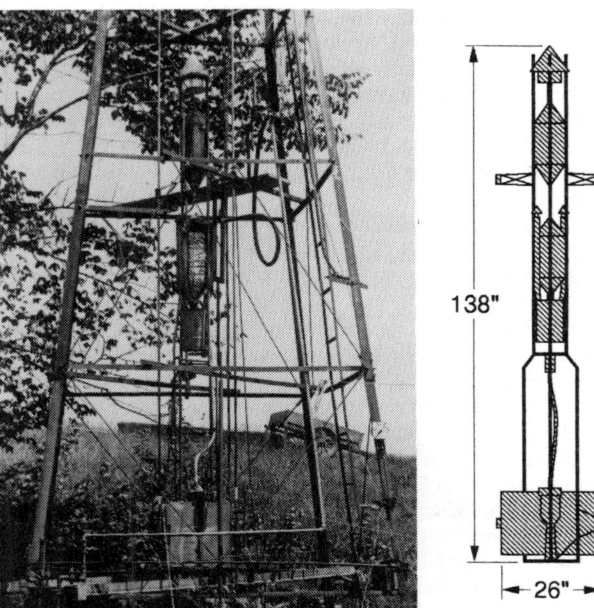

Goddard's rocket of July 17, 1929. (NASA photo 74-H-1071 at left, 1/50 scale sketch by author at right)

Test No. 72: First Flight at Roswell

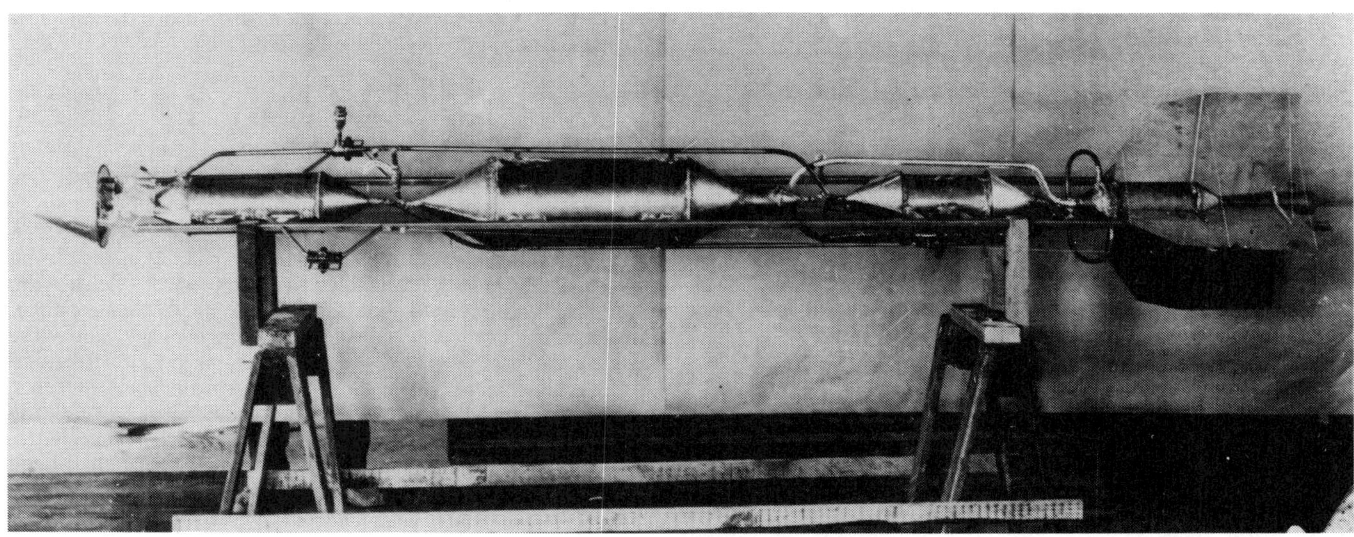

The first Roswell rocket in the shop before its December 30, 1930 flight. Long tube along top of rocket is pressure line. Tube along underside is gasoline line, as are looping pipes entering opposite sides of combustion chamber. Oxygen line runs in a loop around oxygen tank. (NASA photo 74-H-1201)

In his first Roswell tests, Goddard explored means of pressurizing tanks to force propellants into the combustion chamber. He settled on a large tank of pressurized gaseous nitrogen perched between a gasoline tank above and an oxygen tank below. A line from the central nitrogen tank fed into the top of each propellant tank. A valve in the gasoline line controlled flow to the engine, but there was no valve in the oxygen line. Instead, the oxygen line looped over the top of the tank to prevent draining before pressure was applied. The three tanks and the combustion chamber all took the form of cylinders capped by 60° cones—shapes that could be welded entirely from rectangles and semicircles of steel.

This rocket was the last of Goddard's open frame rockets; a conical nose cap and four duralumin (a hard alloy of aluminum) fins were the only concessions to aerodynamics. The loose nose cap concealed a cylindrical parachute compartment. Goddard assumed that the rocket would slide down backwards after reaching apogee. The reverse air flow would pull the cap free, releasing the parachute.

Goddard and his assistants tried to launch the rocket on December 22, 1930, but when they tugged on ropes attached to valves to start the engine, they pulled the rocket from its tower.

On December 30, 1930, Goddard and his crew loaded the rocket into the launch tower, a converted windmill with two guide rails of 1/2" (13 mm) pipe. Four launch shoes, each with three rollers, engaged the pipes. This rocket was painted with a roll pattern—a stripe of bright Chinese red paint down one quadrant. Goddard did not name his rockets (although they picked up the common nickname, Nelle, after Laurence Mansur lamented the previous rocket's bad press with a quote from a popular melodrama—"they ain't done right by our Nelle!"), but he numbered the day's firing "Test No. 72" in his notes.

Goddard and company loaded the gasoline and oxygen tanks, and retreated to a control shelter 50 feet (15 m) from the launcher, and a tracking shelter 1000 feet (300 m) away. Three men—Goddard, Kisk, and Sachs—in the closer hut pulled a series of 10 ropes to seal the oxygen tank from outside air, allow nitrogen pressure in the propellant tanks, to allow gasoline to reach the combustion chamber, to ignite the propellants, and finally to release the rocket. It cleared the tower at about 60 mph (100 km/h), and tilted slightly into the wind. The rocket arced across the sky, reaching 2000 feet (600 m) in 7 seconds, according to the theodolite in the tracking shelter. At the top of the arc, the rocket was still surging forward under the thrust of a small white flame, its nose cone held firmly in place by air resistance. The rocket burned for 8 1/2 seconds and crashed to Earth 1000 feet (300 m) from the launcher. This was by far the highest liquid propelled rocket flight to date, and Goddard set about creating his next generation of rockets.

He carried out a 9-month program of static tests on spherical and cylindrical combustion chambers with an assortment of fuel injection schemes. While Goddard was in the lab, The first liquid-propelled rocket flights were taking place in Germany. Johannes Winkler's HW-1, a "flying test stand," every bit as awkward as Goddard's first rocket, was thought by many at the time to be the first liquid-powered rocket in history, as Goddard had not published his experimental results since 1920 (garbled reports about his 1929 flight led many to believe Goddard's rockets were powered by a series of explosions). The German rocket society, the VfR, followed suit with launches of its Repulsor rockets.

Test No. 72 Specifications

Empty weight without fins	33.5 lb (15 kg)
Thrust	100 lb (450 N)
Duration	8.5 sec
Total Impulse	850 lb-sec (3800 N-sec)
NAR designation	L 450
Altitude	2000 ft (610 m)
Length	11 ft (3.35 m)
Maximum Diameter	9 in (229 mm)

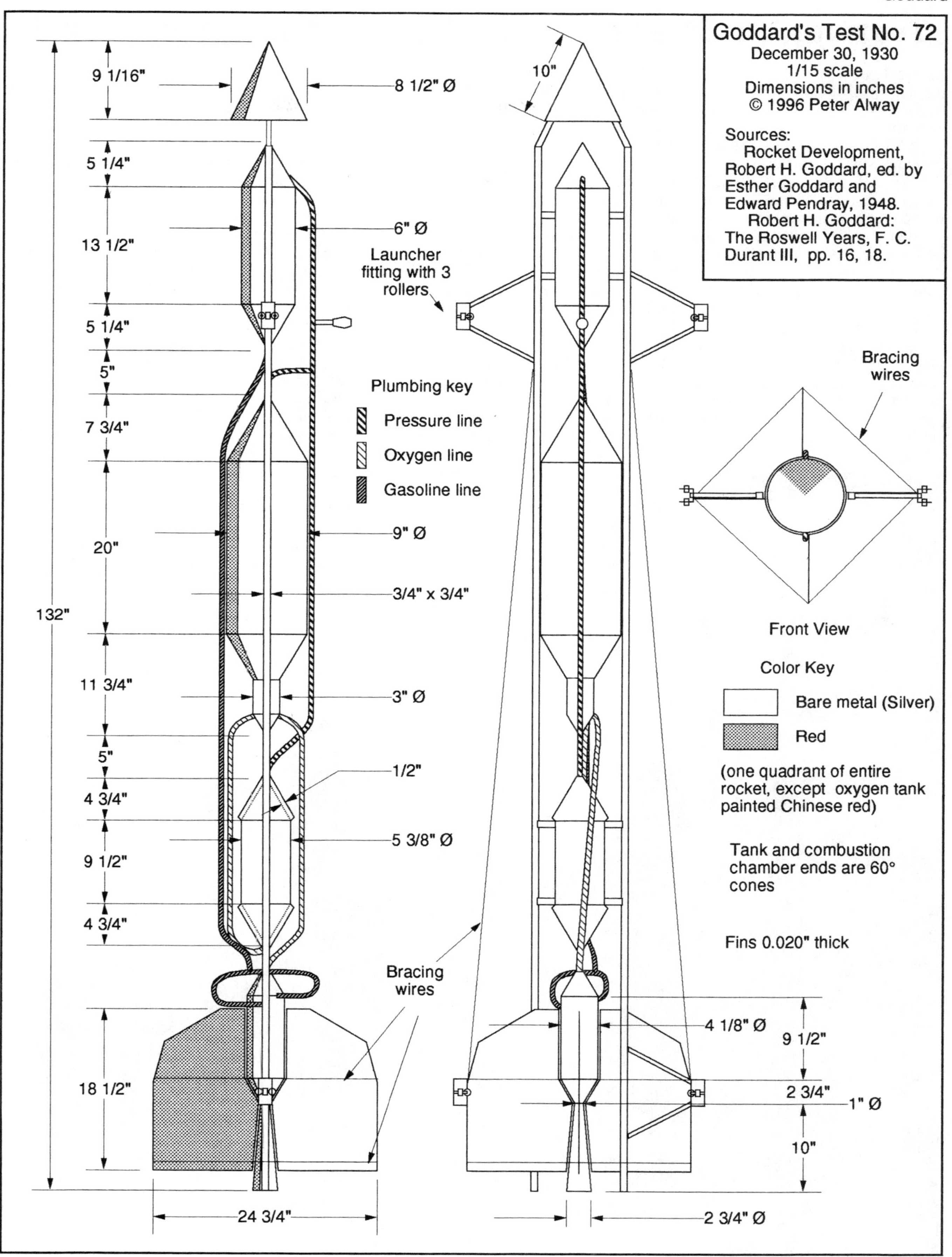
Goddard's Test No. 72
December 30, 1930
1/15 scale
Dimensions in inches
© 1996 Peter Alway
Sources:
Rocket Development, Robert H. Goddard, ed. by Esther Goddard and Edward Pendray, 1948.
Robert H. Goddard: The Roswell Years, F. C. Durant III, pp. 16, 18.
9 1/16"
8 1/2" Ø
10"
5 1/4"
13 1/2"
6" Ø
Launcher fitting with 3 rollers
5 1/4"
5"
7 3/4"
Plumbing key
Pressure line
Oxygen line
Gasoline line
20"
9" Ø
3/4" x 3/4"
132"
11 3/4"
3" Ø
5"
1/2"
4 3/4"
5 3/8" Ø
9 1/2"
4 3/4"
Bracing wires
18 1/2"
24 3/4"
Bracing wires
Front View
Color Key
Bare metal (Silver)
Red
(one quadrant of entire rocket, except oxygen tank painted Chinese red)
Tank and combustion chamber ends are 60° cones
Fins 0.020" thick
4 1/8" Ø
9 1/2"
2 3/4"
1" Ø
10"
2 3/4" Ø

Test No. 73: Streamlined Casing and Automated Launch

In September of 1931, Goddard emerged from the lab with a new flight model. The rocket was powered by a combustion chamber 5 3/4" (146 mm) in diameter—a rocket engine form that would be the basis of Goddard's most successful rockets. In a previous static test, the engine had produced 170 lbs (77 kg) of thrust for 19 seconds. The engine's specific impulse, a measure of efficiency, was 200 seconds. Goddard returned to pressurizing tanks with oxygen, but this time an additional liquid oxygen tank provided tank pressure. Its contents were routed through a jacket of tubes wrapped around the combustion chamber. This eliminated the need for the cumbersome gaseous nitrogen tank of the previous test.

The jacket around the combustion chamber was not designed to cool the combustion chamber—Goddard's method of cooling was to force gasoline to enter the chamber through two holes pointed to set the fuel spiraling around the inner surface of the engine. The liquid fuel would insulate the chamber wall and prevent oxygen from burning away the metal. He also added a protective coating of alundum, a heat-resistant material sold as an abrasive.

A duralumin casing covered the rocket, with the exception of the tank walls. Two lengths of duralumin channel faired over plumbing that extended beyond the tank diameter. A streamlined boattail with fins gave this rocket a more modern look; the launch of this rocket was more modern as well.

On September 29, 1931, Goddard's team trucked the rocket out to the launcher. An electrical command from the control shelter triggered a series of events in a contraption that used weights, ropes, and blasting caps to carry out the sequence of valve turns, gas line releases, and ignition needed to turn on a liquid-propellant rocket.

The automatic ignition sequence was flawless, but the flight wasn't. A yellow flame signaled a shortage of oxygen, and the anemic thrust allowed the rocket to tip to the horizontal after attaining just 180 feet (55 m) of altitude. Deployment of the parachute would be by timer—a device improvised from a child's "Tip-top" watch whose sweep second hand was to close an electrical contact at the expected moment of apogee. Alas, the rocket had crashed well before the appointed moment, damaging the delicate deployment mechanism.

Test No. 73 Specifications

Launch Weight	87.2 lb (39.4 kg)
Empty weight	37 lb (16.8 kg)
Thrust	170 lb (760 N)
Duration	9.6 sec
Total Impulse	1600 lb-sec (7300 N-sec)
NAR designation	M 760
Altitude	180 ft (55 m)
Length	9 ft 11 in (3.02 m)
Diameter	9 3/8 in (238 mm)

Two views of Goddard's rocket of September 29, 1931. Large fairings conceal oxygen, gasoline and pressure lines. (Left, courtesy of Clark University Goddard Library Archives; Right, NASA photo 74-H-1202)

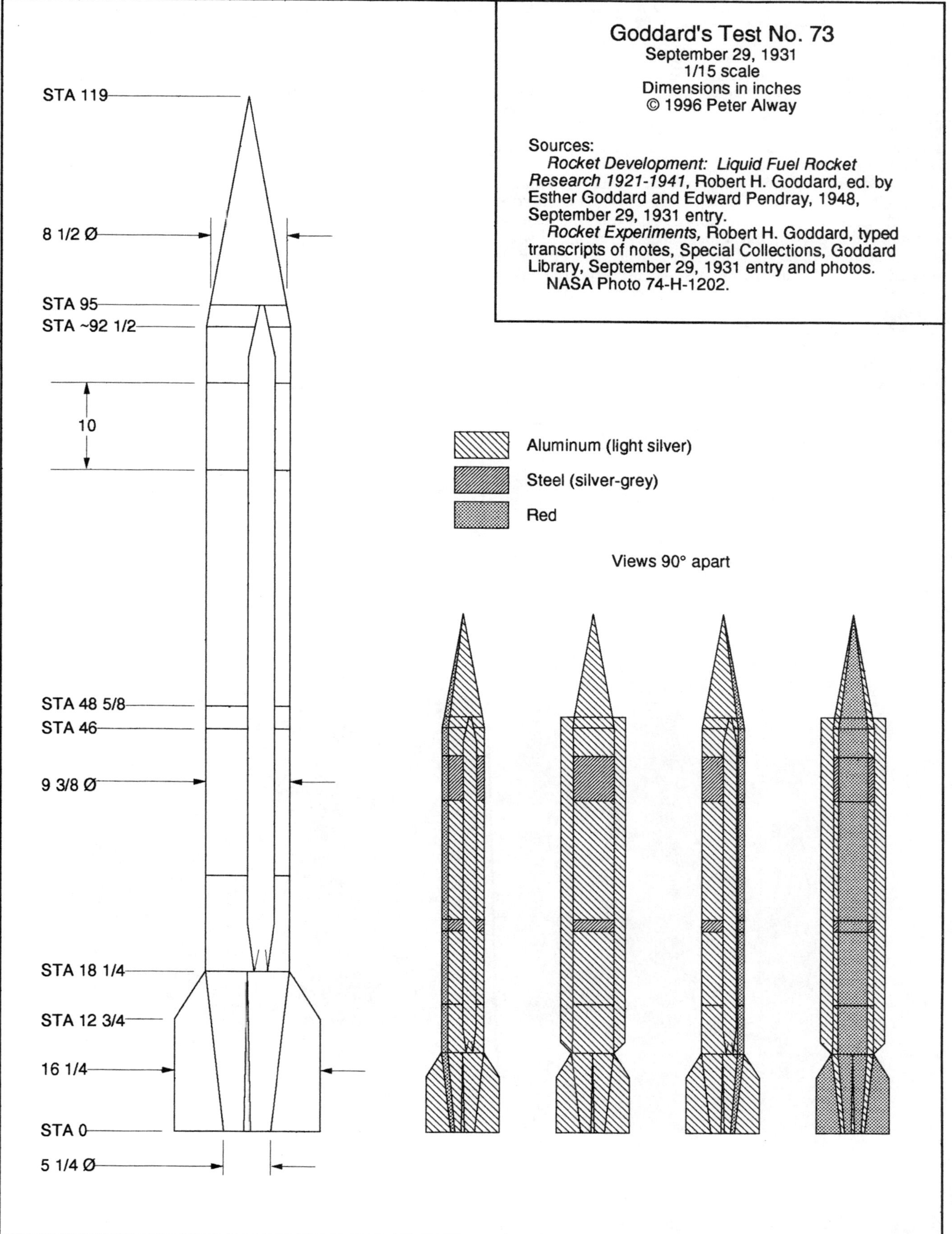
Goddard's Test No. 73
September 29, 1931
1/15 scale
Dimensions in inches
© 1996 Peter Alway
Sources:
Rocket Development: Liquid Fuel Rocket Research 1921-1941, Robert H. Goddard, ed. by Esther Goddard and Edward Pendray, 1948, September 29, 1931 entry.
Rocket Experiments, Robert H. Goddard, typed transcripts of notes, Special Collections, Goddard Library, September 29, 1931 entry and photos.
NASA Photo 74-H-1202.
STA 119
8 1/2 Ø
STA 95
STA ~92 1/2
10
STA 48 5/8
STA 46
9 3/8 Ø
STA 18 1/4
STA 12 3/4
16 1/4
STA 0
5 1/4 Ø
Aluminum (light silver)
Steel (silver-grey)
Red
Views 90° apart

Tests No. 74 and 75: Rockets Bursting in Air

Goddard redesigned the rocket, tweaking the propellant injection, shortening the spaces between tanks, and truncating the long conical nose. In Test No. 74, the new rocket made a flight to 1700 feet (520 m) on October 13, 1931. Soon after engine burnout, as the rocket decelerated, oxygen leaked into the gas tank, and the rocket exploded in mid-air.

Goddard re-built the rocket with remarkable speed—it was ready to fly again in Test No. 75 on October 27, with all new tanks and casings. This time the diameter was reduced to 9", and the conical nose was replaced with a spun aluminum ogive (an ogive is a shape named for the pointed arches of church windows—a curved profile topped by a point). This rocket was left unpainted with no red quadrant to indicate roll. Again the rocket exploded in mid-air, this time at 1330 feet (406 m).

Not only had the October rockets proven the power of the 5 3/4" (146 mm) combustion chamber, but they had pointed out the danger of introducing pure oxygen into the fuel tanks. In the future Goddard would pressurize tanks with nitrogen, sometimes boiled off from a liquid reservoir, and sometimes stored as a pressurized gas.

Test No. 74 Specifications

Thrust	170 lb (760 N)
Duration	13 sec
Total Impulse	2200 lb-sec(9800 N-sec)
NAR designation	M 760
Altitude	1700 ft (520 m)
Length	7 ft 9 in (2.36 m)
Diameter	9 3/8 in (238 mm)

Test No. 75 Specifications

Empty weight	39.5 lb (18 kg)
Thrust	170 lb (760 N)
Duration	9.6 sec
Total Impulse	1600 lb-sec (7300 N-sec)
NAR designation	M 760
Altitude	1330 ft (406 m)
Length	8 ft 7 in (2.62 m)
Diameter	9 in (229 mm)

Goddard and his crew prepare to load the rocket of October 17, 1931 into the launch tower. (NASA photo 74-H-1195)

Goddard's Rockets of October, 1931

1/15 scale
Dimensions in inches

Sources:

Rocket Development: Liquid Fuel Rocket Research 1921-1941, Robert H. Goddard, ed. by Esther Goddard and Edward Pendray, 1948, entries for May 23, September 29, October 13, and October 27, 1931.

The Papers of Robert H. Goddard, Ed. by Esther C. Goddard, McGraw-Hill, 1970, pp. 815, 1160.

Rocket Experiments, Robert H. Goddard, typed transcripts of notes, Special Collections, Goddard Library, entries for September 29, October 13, and October 27, 1931.

NASA Photo 74-H-1195.

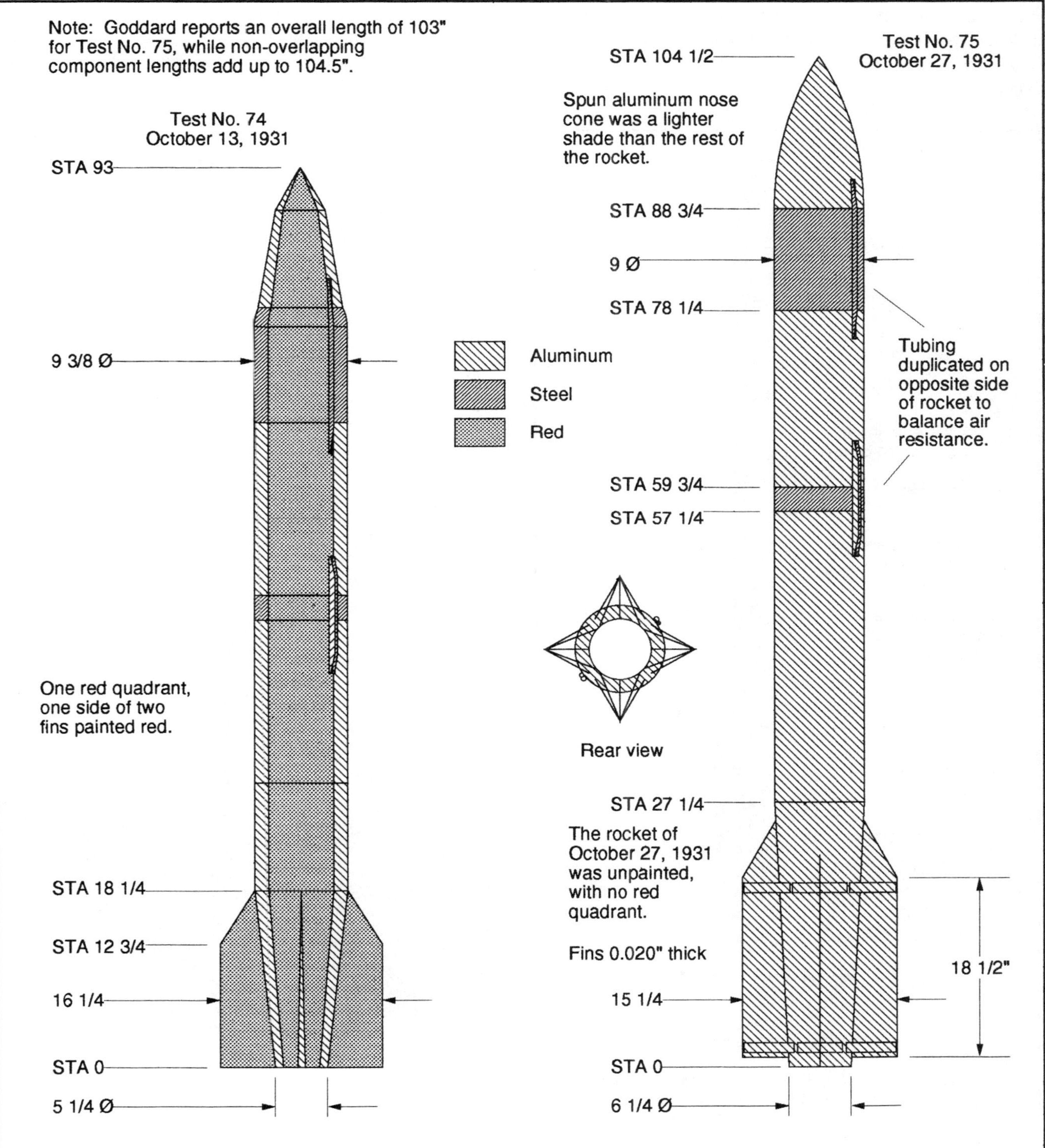

Test No. 77: The First Gyroscope-Stabilized Rocket

Goddard finished his first Roswell campaign with the launch of a gyroscope stabilized rocket. The 4" (10 cm)-diameter gyroscope was mounted on gimbals in the rocket. If the rocket deviated from vertical, it would close electrical contacts on the gimbals. Current through these contacts would trigger small bell magnets which would open pneumatic valves. Gas pressure would then move small vanes into the rocket exhaust as needed to right the rocket. Small air vanes were attached outside the blast vanes to allow guidance after the propellants were exhausted.

STA ~149 1/2"
STA 129 1/2"
12" Ø
Aluminum (Light Silver)
Red
Control vanes are steel (dark silver)
One red quadrant, one side each of 2 fins red
STA 29 3/4"
45°
3/4" maximum fin thickness
18"
STA 3 1/4"
STA 0
20 1/2"
8 1/2" Ø
1"

Goddard Test No. 77
April 19, 1932
1/30 scale
©1996 Peter Alway

Sources:
Rocket Experiments, Robert H. Goddard, typed transcripts of notes, Special Collections, Goddard Library, entries for November 18, 1931-April 19, 1932.
NASA Photo 74-H-1203.

A bellows pump forced liquid nitrogen from a supply tank through a jacket of tubing wrapped around the engine. The evaporated gas provided pressure both for the control system and to force propellants into the combustion chamber. The 9" (229 mm) propellant tanks were concealed entirely within a 12" (305 mm) casing.

The rocket flew April 19, 1932, and promptly tipped to one side. The vanes did not right the rocket, but immediately after the flight, Goddard felt each of the blast vanes. The one that should have corrected the path of the rocket was warm, but the vanes had been too small to keep the rocket pointed straight up.

Goddard was out of money, and in the depths of the Depression, even his wealthy patrons in the Guggenheim family (Daniel Guggenheim had died in 1930) withdrew support. Goddard returned to Clark University to teach and to conduct what research he could on details of rocket construction.

Test No. 77 Specifications

Empty weight	91.5 lb (41.5 kg
Duration	5 sec
Altitude	135 ft (40 m)
Length	10 ft 9 1/2 in (3.29 m)
Diameter	12 in (305 mm)

The first Gyroscope-stabilized rocket, launched on April 19, 1932. (NASA photo 74-H-1203)

Interlude at Clark University

In 1931, Goddard had applied for a patent on a "rocket chamber utilizing atmospheric air." Hearkening back to his multiple charge solid rockets, this device used a resonating series of gasoline explosions to produce thrust. Air was admitted through the front of the chamber to mix with gasoline. When the combination exploded in the presence of a spark plug, shutters at the front of the chamber closed, forcing exhaust gasses out the rear. From November of 1932 to June of 1933, the professor tinkered with this invention. In one test, Goddard attached a small model to the rim of a bicycle wheel to permit forward motion to replenish the air supply. In another test, the device sustained thrust at two explosions per second.

Unknown to Goddard, the German aviation magazine *Flugsport* published a translation of his patent in its January, 1939 issue. The patent outlined the principles of the pulse-jet engine that would give the V-1 "Buzz Bomb" its buzz in World War II.

Goddard also tinkered with propellant pumps at Clark. He abandoned bellows pumps and recognized that only a turbine-powered centrifugal pump would provide the flow rate and pressure he required. Even today, the Space Shuttle relies on such pumps. He tested small pumps, first with water, then gasoline, and finally liquid oxygen.

While Robert Goddard was working in Massachusetts, the German VfR broke up. Wernher von Braun flew his sausage-shaped, passive-gyro stabilized A-2 rockets for the German Army. The German Army quietly began hiring the the most talented of the VfR's experimenters to work on a ballistic missile that would come to be known as the V-2. In the Soviet Union, Sergei Korolev's GIRD group had launched the hybrid GIRD-09 rocket and the liquid-powered GIRD-X. Even the amateur American Interplanetary Society managed to fly a liquid rocket from Staten Island. Goddard was beginning to lose his lead in rocket development. In 1934, the Guggenheims resumed their support, and Goddard resumed his work at Roswell.

The A-Series

Goddard chose to pursue high altitude flight with a series of lightweight, simple rockets burning gasoline and liquid oxygen forced into the combustion chamber by the pressure of nitrogen gas. Typical "A-series" rockets were 9" (229 mm) in diameter and and around 15 feet (4.5 m) long. Each was capped by a blunt ogive nose cone and had four delta fins bolted to a conical boattail. Internally, these rockets followed the form of the first Roswell rocket. Behind an aluminum nose cone was the gasoline tank. On most flights the parachute was stowed in a box attached to the rocket's skin behind the gasoline tank. The nitrogen pressure tank was next, constituting almost a third of the rocket's length. The nitrogen not only forced propellants into the engine, but it powered the gyroscope and steering vanes. Behind it were the oxygen tank, and finally the tail section, concealing the engine. Steering vanes, designed to flip into the rocket exhaust to deflect it as needed, trailed behind all but one of the A-series rockets.

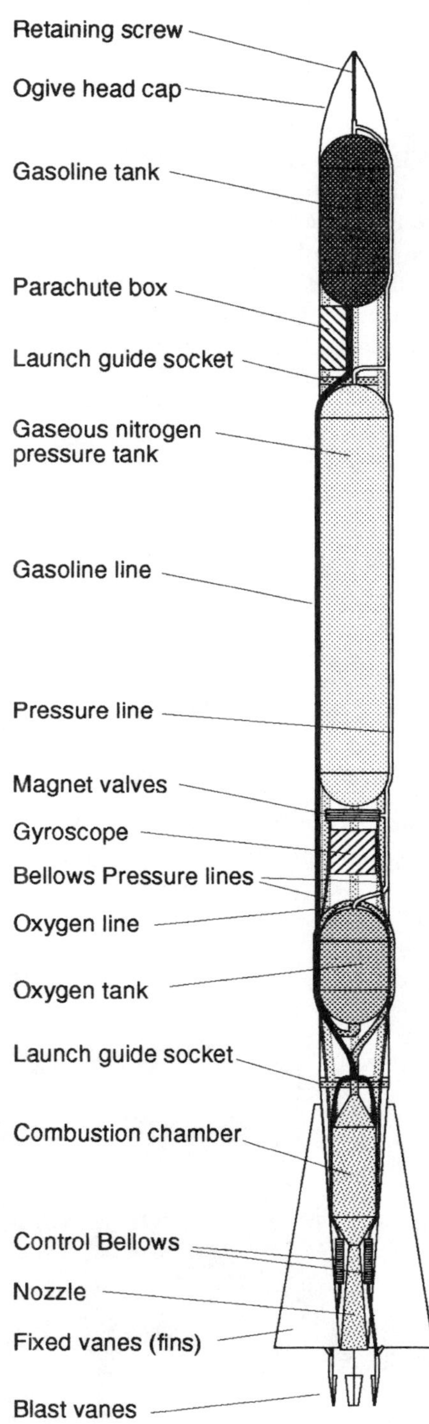

Components of a typical A-series rocket. Details of the internal arrangement varied from flight to flight. Some components are rotated into the plane of the drawing for illustration. (drawing by author)

Rocket A-3

The first A-series rockets, A-1 and A-2, failed to leave the ground, forcing Goddard to simplify the design, removing the guidance and control mechanisms. He also replaced the original rocket's cylindrical tail and quadrilateral fins with a conical boattail and delta fins.

Goddard's team trucked out the new incarnation of the rocket, A-3, to the launch tower on February 16, 1935. After loading the rocket into the tower, Robert Goddard took his place at the launch control keys. A gust of wind had broken the keyboard support, so Al Kisk held it for the professor. With three telegraph keystrokes, Goddard pressurized the tanks, ignited the rocket, and set it free.

Esther Goddard followed the rising machine with a movie camera. Charles Mansur (Laurence Mansur's brother) followed the rocket with a recording telescope. Nils Ljundquist timed the flight.

The rocket accelerated from the tower at about 2 g's. But the rocket engine was fed an excess of oxygen, which attacked the wall of the combustion chamber. The rocket let out an audible pop, and flame shot out of the side of the combustion chamber. The rocket reached just 220 feet (67 m). The parachute opened, and cushioned the rocket's fall. Much of the rocket would fly again, although Goddard would have to concoct a new oxygen pressure regulator if the next rocket were to perform better.

A-3 Specifications

Empty weight	58 lb (26 kg)
Thrust	193 lb (860 n)
Altitude	220 ft (67 m)
Length	13 ft 6 3/8 in (4.12 m)
Diameter	9 in (229 mm)

The first A-series rocket flight was test A-3, a simplified model without a guidance system. Throughout the A-series, Goddard repaired and re-used rockets as much as possible—whether the first A-series tests were three flights of one rocket or three rockets of one type is a matter of semantics. (courtesy of Clark University Goddard Library Archives)

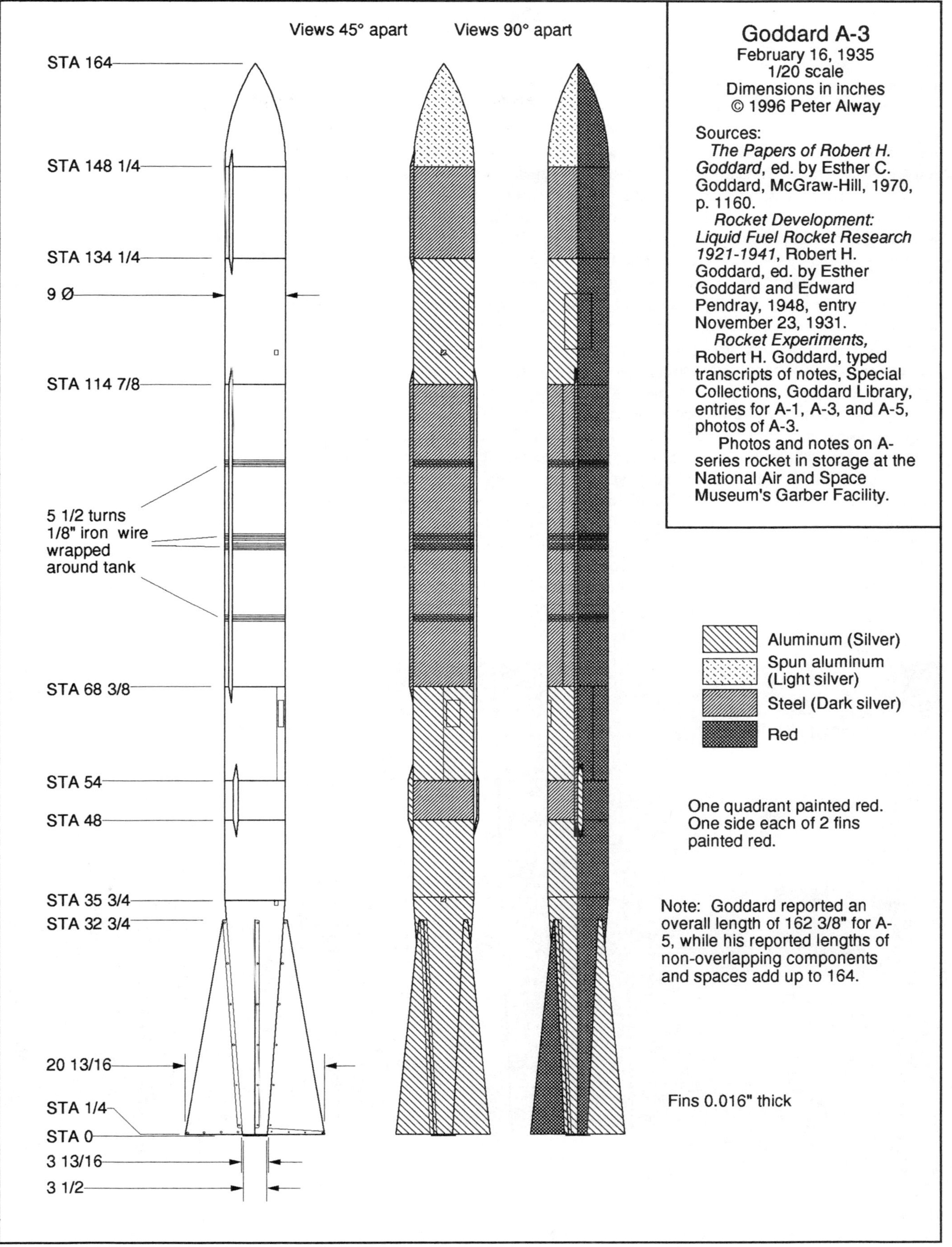

Goddard A-3

February 16, 1935
1/20 scale
Dimensions in inches

Sources:

The Papers of Robert H. Goddard, ed. by Esther C. Goddard, McGraw-Hill, 1970, p. 1160.

Rocket Development: Liquid Fuel Rocket Research 1921-1941, Robert H. Goddard, ed. by Esther Goddard and Edward Pendray, 1948, entry November 23, 1931.

Rocket Experiments, Robert H. Goddard, typed transcripts of notes, Special Collections, Goddard Library, entries for A-1, A-3, and A-5, photos of A-3.

Photos and notes on A-series rocket in storage at the National Air and Space Museum's Garber Facility.

Rockets A-4 and A-5

Goddard chose not to risk his fragile and expensive gyroscope (one of the few commercially available components of his rockets) on the first test of a new control system. Instead he equipped rocket A-4 with a pendulum to actuate the thrust control vanes (the significance of the pendulum system in Goddard's work has been exaggerated by some—this was simply a one-flight substitute for a gyro). The pendulum would maintain its orientation in the ascending rocket only briefly, but long enough to see if the control vanes functioned. The rocket lifted off on March 8, 1935. As expected, the control system corrected the rocket's path for a few seconds, but as the rocket approached 1000 feet (300 m), it took to horizontal flight. Roaring above the desert, it ripped free from its parachute and attained what was probably the highest speed to date for a human invention flying under its own power. The tracking theodolite was connected to a pair of clockwork chart recorders, and from this record, Goddard determined that this rocket had probably exceeded the speed of sound.

On March 28, a new rocket (new can be a misleading term, as the thrifty Robert Goddard re-used as many components as he could from flight to flight), A-5, was ready to fly. This model carried a fully functional gyroscope, designed to close electrical contacts when the rocket deviated 10° from the vertical. The closed contacts in turn sent electric currents through coils in magnetic valves that let pressurized gas into bellows that would force the appropriate blast vanes into the rocket flame.

As the rocket rose, Esther Goddard noted that the rocket flew "like a fish swimming upwards through water." The guidance system worked, maintaining the rocket on a vertical course to an altitude of 4800 feet (1460 m). As the rocket approached the apex of its flight, Kisk noted flashes of red paint—the rocket was spinning about its axis about twice a second. Perhaps a thrust vane had been warped by the rocket's flame. The parachute failed to operate; centrifugal effects inside the spinning rocket may have jammed the mechanism. Nonetheless, Goddard and three assistants proudly posed with the wreckage for Esther's camera. For the first time, a rocket had operated under the control of an active inertial guidance system.

A-5 Specifications

Launch Weight	127 lb (58 kg)
Empty weight	78.5 lb (36 kg)
Thrust	193 lb (860 N)
Duration	14 sec
Total Impulse	2700 lb-sec (12,000 N-sec)
NAR designation	N 860
Altitude	4800 ft (1460 m)
Length	14 ft 9 3/4 in (4.51 m)
Diameter	9 in (229 mm)

Robert H. Goddard poses with rocket A-5, the first rocket to fly successfully with an inertial guidance system, including a gyroscope controlling blast vanes extending from the rear of the rocket. (courtesy of Clark University Goddard Library Archives)

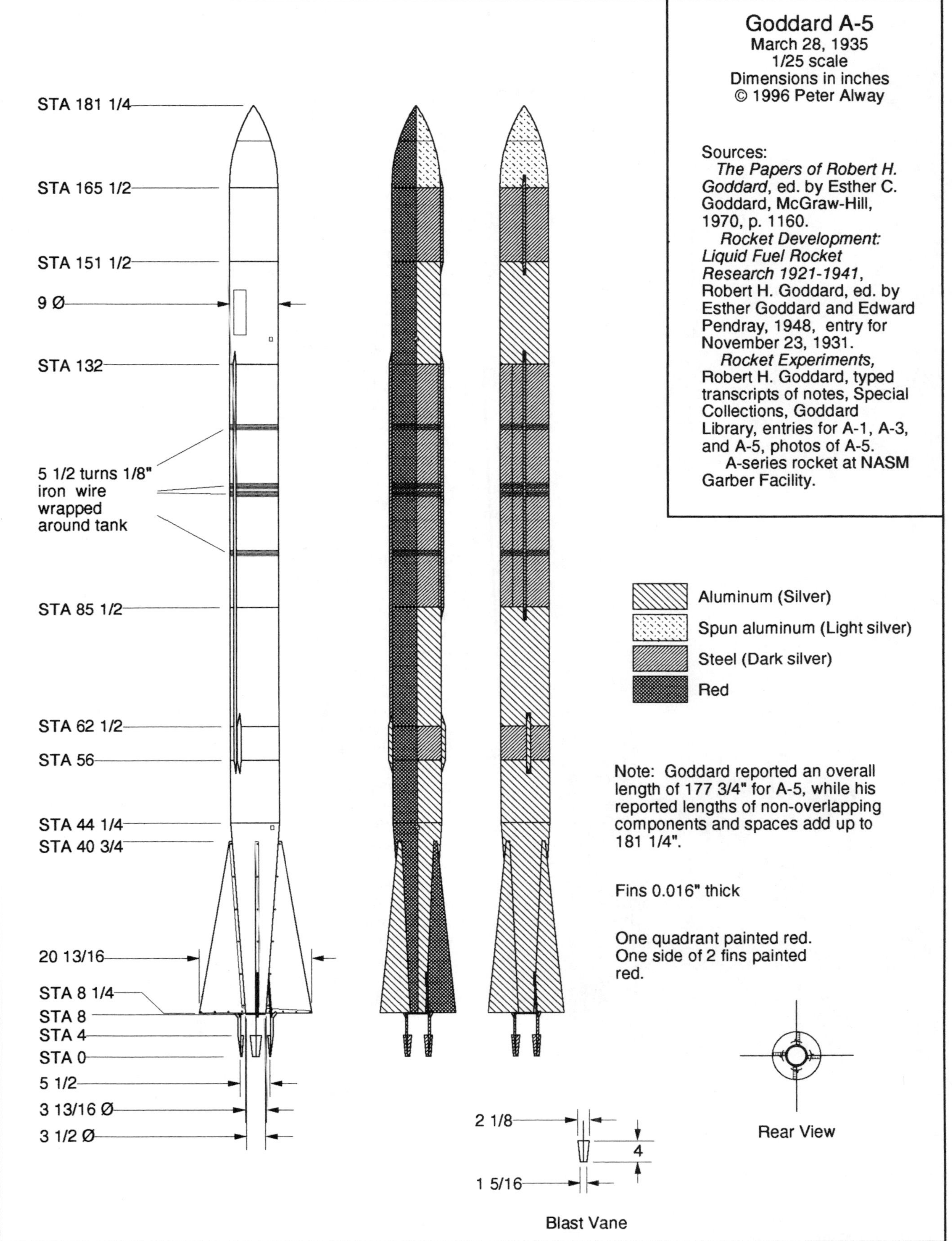
Goddard A-5
March 28, 1935
1/25 scale
Dimensions in inches
© 1996 Peter Alway
Sources:
The Papers of Robert H. Goddard, ed. by Esther C. Goddard, McGraw-Hill, 1970, p. 1160.
Rocket Development: Liquid Fuel Rocket Research 1921-1941, Robert H. Goddard, ed. by Esther Goddard and Edward Pendray, 1948, entry for November 23, 1931.
Rocket Experiments, Robert H. Goddard, typed transcripts of notes, Special Collections, Goddard Library, entries for A-1, A-3, and A-5, photos of A-5.
A-series rocket at NASM Garber Facility.
STA 181 1/4
STA 165 1/2
STA 151 1/2
9 Ø
STA 132
5 1/2 turns 1/8" iron wire wrapped around tank
STA 85 1/2
STA 62 1/2
STA 56
STA 44 1/4
STA 40 3/4
20 13/16
STA 8 1/4
STA 8
STA 4
STA 0
5 1/2
3 13/16 Ø
3 1/2 Ø
Aluminum (Silver)
Spun aluminum (Light silver)
Steel (Dark silver)
Red
Note: Goddard reported an overall length of 177 3/4" for A-5, while his reported lengths of non-overlapping components and spaces add up to 181 1/4".
Fins 0.016" thick
One quadrant painted red. One side of 2 fins painted red.
2 1/8
4
1 5/16
Blast Vane
Rear View

Rocket A-14, the last A-rocket to fly, in the launch tower. The tower was a converted windmill. The rocket's red quadrant always faced the rear of the shelter. The tower supported two vertical pipes to guide the rockets during launch. A-series rockets engaged the pipes with rollers that sprung free of the rocket at launch. (NASA photo 74-H-1219)

The Final A-Series Rockets

Flights A-8, A-9, and A-10 followed. Two of these, A-8 and A-10, reached altitudes of over a mile. With a series of successes under his belt, Goddard invited Lindbergh and Harry Guggenheim, Daniel's heir, to Roswell to witness a launch.

On September 23, Goddard's crew loaded up rocket A-11 to demonstrate to the men who had the influence and money to keep American rocketry in the forefront. But on ignition, the rocket sputtered with an anemic yellow flame, and refused to move. No stranger to failure, Goddard had a second rocket in store. After a day of rain, Rocket A-12 was ready to fly on September 25. Again, the rocket refused to lift.

Lindbergh and Guggenheim left, disappointed in not seeing a flight, but relieved to see the program advancing in a professional manner. Over the next few weeks Goddard found the cause of the problem. One of his assistants, usually impeccable craftsmen, had drilled a fuel injection hole in the combustion chamber with the wrong bit. The rocket from test A-12 was repaired, and launched in test A-14. In the absence of important guests, it flew to 4000 feet (1220 m) on October 29, 1935. This was the last of the A-series tests, but when Goddard ran into dead ends in later work, he would frequently return to the ideas and components that worked in the A-series.

At the urging of Lindbergh, Goddard pieced together one complete A-series rocket (as A-14 crashed under power, this would probably be the rocket left from test A-11), cleaned it, greased it, oiled it, and shipped it off to the Smithsonian, where, at the professor's insistence, it would be sealed behind a false wall until it was so obsolete as to be of no use to his competitors. After Goddard's death, it was displayed at the museum (the rocket was eventually displaced by higher-flying spaceships—at the time of writing, this rocket lies secured in a plexiglass sarcophagus stuffed behind a Stearman biplane at the Smithsonian's Garber Facility). Goddard's insistence on secrecy had proved an annoyance to his colleagues in rocketry. Goddard was especially worried about rocket development in Germany. While Von Braun might have benefited from a peek at the type A rocket, his American colleagues would have benefited more. Fortunately, Lindbergh also persuaded Dr. Goddard to published *Liquid-Propellant Rocket Development* through the Smithsonian Institution. While not complete and up to date, this book gave the outside world its first undistorted view of Goddard's work since 1920.

Goddard's greater goal was to develop the full set of components—propulsion, guidance, and control—for a space rocket, and to demonstrate them in an atmospheric rocket. To Goddard, progress was measured in the refinement of each part. But to the outside world, and even to his ever-patient patrons, progress was measured in altitude. Higher altitude would require larger rockets, and so began the K-series tests.

A-14 Specifications

Empty weight	84.5 lb (38 kg)
Duration	14 sec
Altitude	4000 ft (1220 m)
Length	15 ft 3 1/2 in (4.66 m)
Diameter	9 in (229 mm)

A single A-series rocket remains in existence, at the National Air and Space Museum's Garber facility, stored behind a biplane of the same era. This rocket, or most of it, is probably the vehicle that refused to leave the ground in test A-11, conducted in the presence of Charles Lindbergh and Harry Guggenheim. Goddard himself certified only that the major components of this rocket, with the exception of the propellant tanks, and flown. It differs from the earlier A-series rockets in several refinements, including thicker fins and small aerodynamic control vanes attached to the blast vane supports. (photos by author)

L-5 and L-6: Large Combustion Chambers

The K-series was a group of static tests of a large rocket engine 10" (254 mm) in diameter. By May of 1936, Goddard was ready to incorporate the combustion chambers into a new rocket design, called the L-series. The first L-rockets eliminated the need for a huge nitrogen tank by carrying liquid nitrogen as the pressurant. L-5 was the first of these to fly, on July 31, 1935.

L-5 was Zeppelin-shaped machine, shorter than the preceding A-series, but 18" (457 mm) in diameter to accommodate the larger engine. In addition to four blast vanes around the nozzle, there were four aerodynamic vanes trailing behind the fin tips. The flight ended at an altitude of 200 feet (60 m), when the combustion chamber burned through. A second attempt with a similar rocket, test L-6, on October 3, suffered the same fate.

Cooling a rocket combustion chamber was one of the greatest challenges that 1930's rocketeers faced. Eventually, regenerative cooling, in which fuel was forced through tubes in the engine wall before burning, would emerge as the method of choice. Goddard, on the other hand, was dedicated to a method called "curtain cooling," In which a layer of gasoline inside the combustion chamber would prevent a burn-through. With the failures of the first two L-series tests, it was clear that the professor could not yet scale the technique up to the larger engines.

L-5 Specifications

Launch Weight	297 lb (135 kg)
Empty weight	135 lb (61 kg)
Thrust	550 lb (2400 N)
Altitude	200 ft (60 m)
Length	12 ft 10 in (3.91 m)
Diameter	18 in (457 mm)

L-6 Specifications

Launch Weight	241.5 lb (110 kg)
Empty weight	151.49 lb (68.8 kg)
Thrust	550 lb (2400 N)
Duration	5 sec
Total Impulse	2800 lb-sec (12,000 N-sec)
NAR designation	N 2400
Altitude	200 ft (60 m)
Length	13 ft 4 1/2 in (4.08 m)
Diameter	18 in (457 mm)

Goddard's two attempts to fly rockets with large combustion chambers. Above is the rocket that would fly as test L-5, (photographed before an earlier static test). The gasoline tank is exposed at the front of the 18" casing. Note blast vanes close to the nozzle, and movable air vanes near the fin tips. Below is rocket L-6, whose tanks and piping were entirely enclosed in its 18" casing. (courtesy of Clark University Goddard Library Archives)

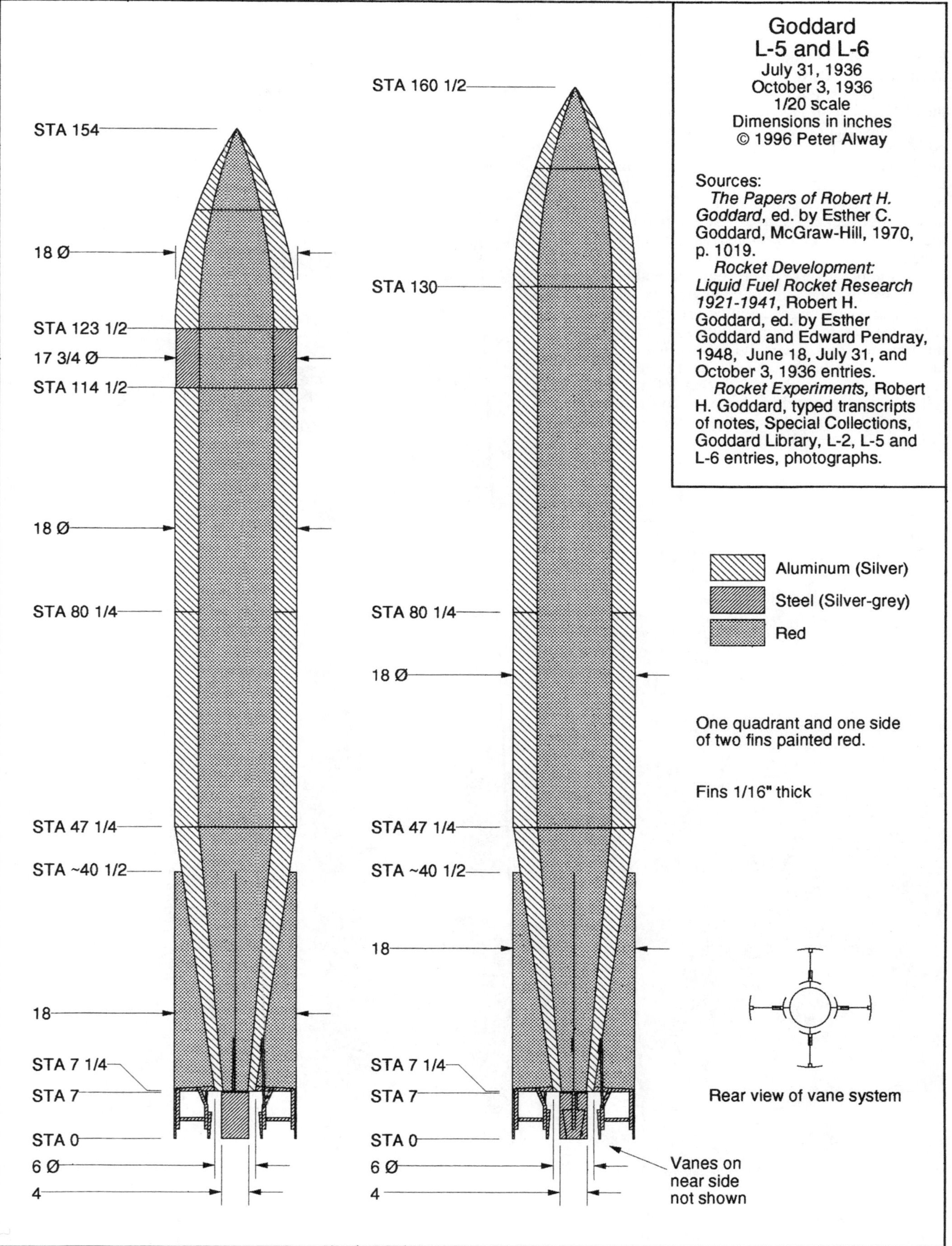
Goddard
L-5 and L-6
July 31, 1936
October 3, 1936
1/20 scale
Dimensions in inches
© 1996 Peter Alway
Sources:
The Papers of Robert H. Goddard, ed. by Esther C. Goddard, McGraw-Hill, 1970, p. 1019.
Rocket Development: Liquid Fuel Rocket Research 1921-1941, Robert H. Goddard, ed. by Esther Goddard and Edward Pendray, 1948, June 18, July 31, and October 3, 1936 entries.
Rocket Experiments, Robert H. Goddard, typed transcripts of notes, Special Collections, Goddard Library, L-2, L-5 and L-6 entries, photographs.
STA 154
18 Ø
STA 123 1/2
17 3/4 Ø
STA 114 1/2
18 Ø
STA 80 1/4
STA 47 1/4
STA ~40 1/2
18
STA 7 1/4
STA 7
STA 0
6 Ø
4
STA 160 1/2
STA 130
STA 80 1/4
18 Ø
STA 47 1/4
STA ~40 1/2
18
STA 7 1/4
STA 7
STA 0
6 Ø
4
Vanes on near side not shown
Aluminum (Silver)
Steel (Silver-grey)
Red
One quadrant and one side of two fins painted red.
Fins 1/16" thick
Rear view of vane system

L-7: Four-Chamber Rocket

Rather than abandon the larger rocket attempts, Goddard returned to his earlier A-type engines, but mounted four of them into the airframe of an 18" (457 mm) rocket, L-7.

The first clustered liquid rocket was an odd-looking affair—Goddard refused to give up the aerodynamic advantage of a boattail, so the four engines resided in cylinders protruding from the rear of a 60° cone. As before, the rocket sported four fixed fins, but the four blast vanes doubled as aerodynamic vanes.

Launch took place on November 7, 1936. While each individual engine was more reliable than its larger counterpart, there were four times as many opportunities to fail. By the time the rocket left the tower, one engine had burned through. Again, an 18" monster rocket crept up to about 200 feet (60 m) and fell to Earth. After three large rockets refused to approach extreme altitudes, it was time to think small again.

L-7 Specifications

Empty weight	202 lb
Length	13 ft 6 1/2 in (4.13 m)
Diameter	18 in (457 mm)

One of Goddard's crew inspects the four chambers and control vanes of Rocket L-7. (courtesy of Clark University Goddard Library Archives)

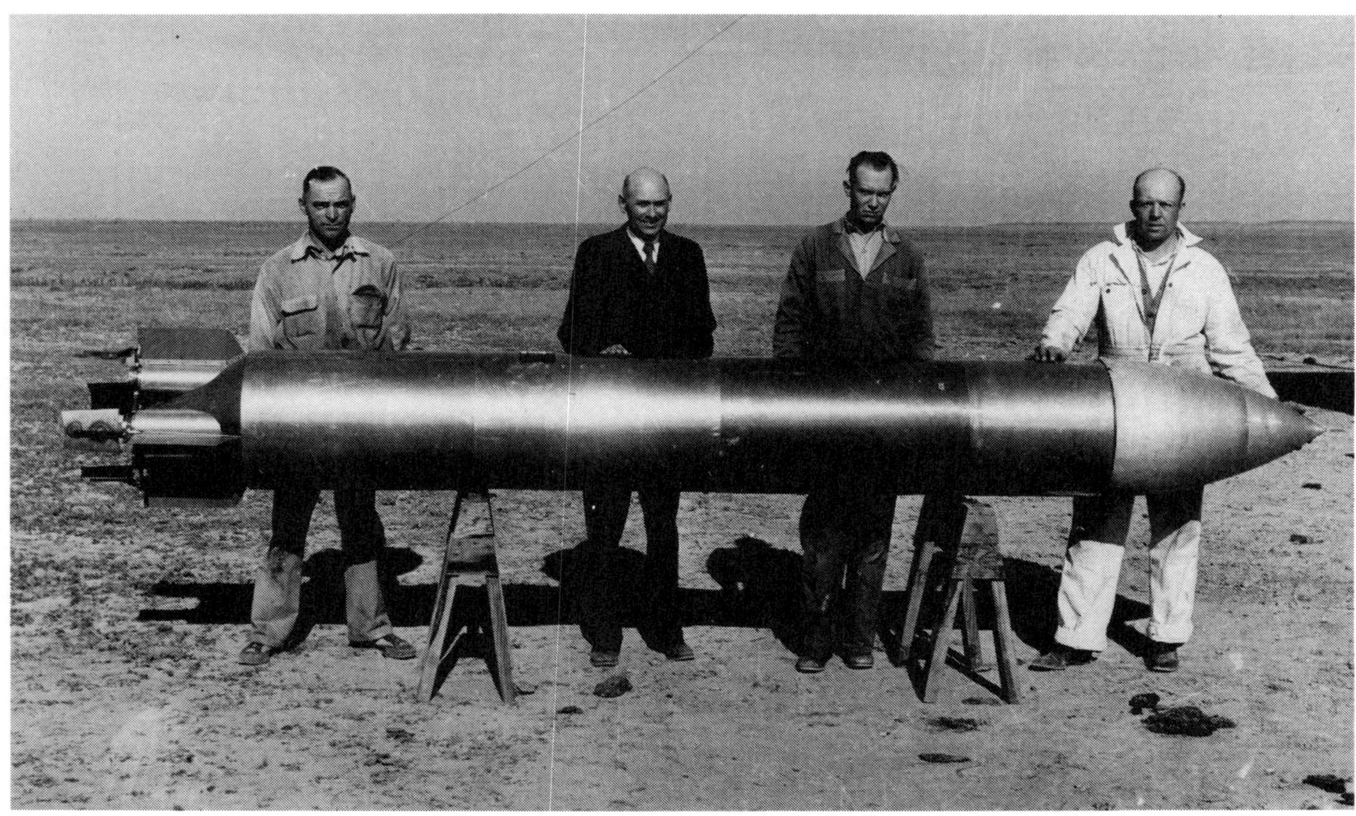

Goddard's team lines up with the four-chambered cluster rocket L-7. (Courtesy Goddard Library Special collections. This photograph is available as NASA 74-H-1226)

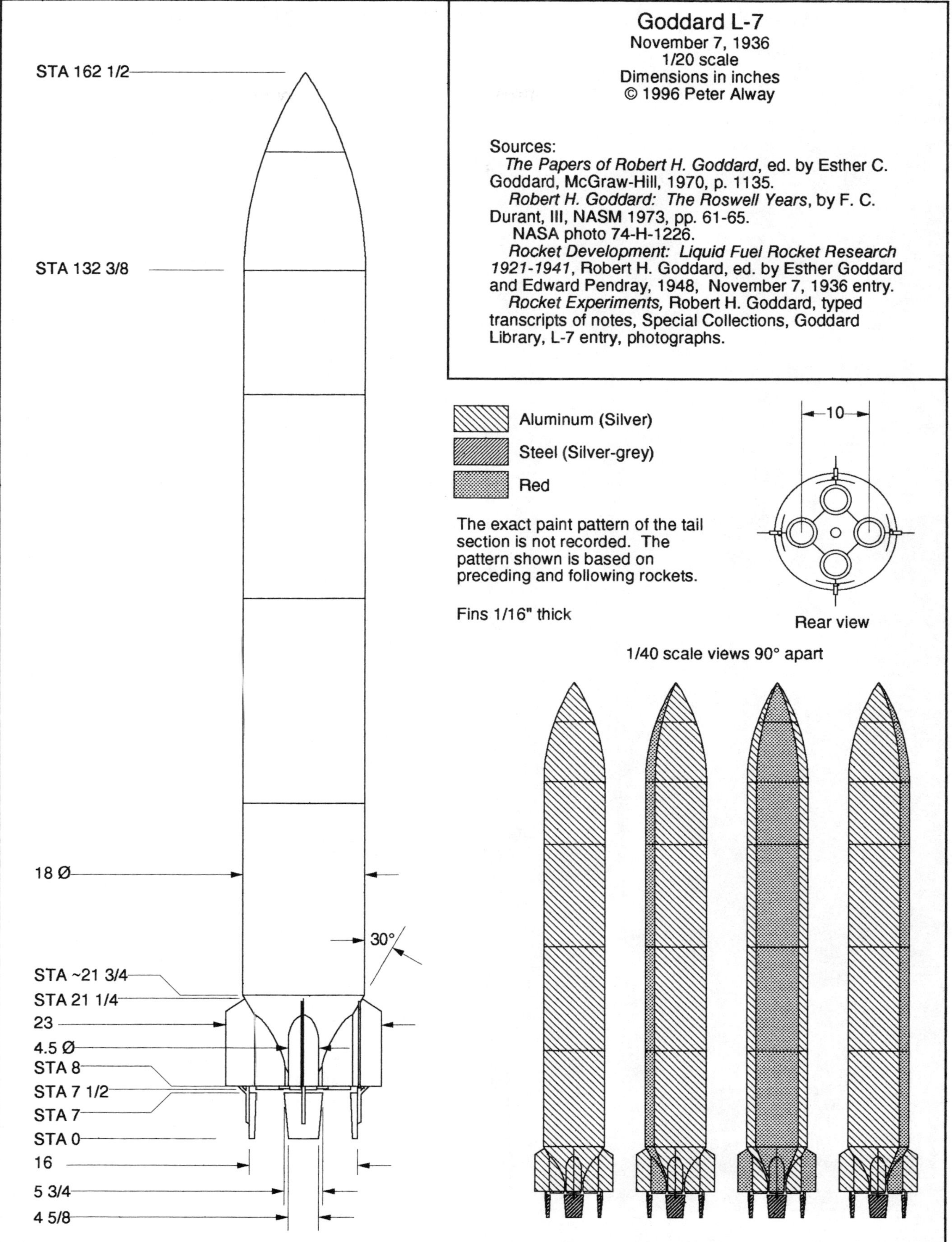

Goddard L-7
November 7, 1936
1/20 scale
Dimensions in inches
© 1996 Peter Alway
Sources:
The Papers of Robert H. Goddard, ed. by Esther C. Goddard, McGraw-Hill, 1970, p. 1135.
Robert H. Goddard: The Roswell Years, by F. C. Durant, III, NASM 1973, pp. 61-65.
NASA photo 74-H-1226.
Rocket Development: Liquid Fuel Rocket Research 1921-1941, Robert H. Goddard, ed. by Esther Goddard and Edward Pendray, 1948, November 7, 1936 entry.
Rocket Experiments, Robert H. Goddard, typed transcripts of notes, Special Collections, Goddard Library, L-7 entry, photographs.
Aluminum (Silver)
Steel (Silver-grey)
Red
The exact paint pattern of the tail section is not recorded. The pattern shown is based on preceding and following rockets.
Fins 1/16" thick
10
Rear view
1/40 scale views 90° apart
STA 162 1/2
STA 132 3/8
18 Ø
30°
STA ~21 3/4
STA 21 1/4
23
4.5 Ø
STA 8
STA 7 1/2
STA 7
STA 0
16
5 3/4
4 5/8

L-10 through L-15: A Return to Smaller Rockets

After the failures of the large rockets, Goddard returned to a design much like the A-series rockets—machines 9" (229 mm) in diameter and relying on gaseous nitrogen to force propellants into a 5 3/4" (146 mm) combustion chamber. Before resuming flights, the professor tested automotive gasoline, aviation gasoline, and naptha in a series of static tests numbered L-8 and L-9.

Goddard explored various configurations of blast vanes, aerodynamic control vanes, and fixed fins on flights L-10 through L-15.

Rocket L-10 (photo on page 4) was unguided, stabilized only by four fins a bit smaller than those of the A-series rockets. For purposes of streamlining, gasoline and pressure lines ran through the nitrogen pressure tank. L-10 lifted off on December 18, 1936 with a roar that echoed across eight miles of desert to the Goddards' ranch house. This model quickly turned horizontal at launch, and crashed 2000 feet from the tower.

L-11 carried a gyroscope and movable air and blast vanes similar to those of L-6. On February 1, 1937, L-11 shot from the tower faster than any of Goddard's previous tests, and maintaining its attitude 10° from vertical, flew to 1870 feet. The flight was marred by a failure of the timed parachute system. The next rocket, L-12, flew on February 27. While the guidance system struggled to keep the rocket upright, it tilted progressively into the wind. After reaching an altitude of 1500 feet (460 m), it crashed into the desert floor with 10 seconds of fuel remaining. Goddard saw a mushroom-shaped cloud rise over the crash site.

Goddard's crew took rocket L-13 out to the tower at noon on March 26, 1937. Wind gusts, remnants of the previous day's dust storm, promised to delay the launch, so the crew mounted an improvised windsock—a scrap of cloth—on the launch tower. Goddard, hovering over the master control board's telegraph keys, watched the cloth for signs of a lull that would allow a safe flight. At 4:10, the professor found a moment of good air, and pressed the keys to start the engine and release the rocket.

Just as the rocket cleared the tower, the wind returned, blowing the rocket's fins downwind, and cocking the nose upwind. But Nelle's electromechanical brain sensed the tilt and corrected, twitching her control vanes in response. Charles Mansur followed with the recording telescope while Nelle chugged out smoke, but within seconds, the engine began to burn more cleanly, and Mansur lost the silver rocket in the dusty sky. Nils Ljundquist timed a 22.3 second burn with his stopwatch. Al Kisk at the theodolite was able to follow the rocket to apogee, but his instrument wasn't. His estimated elevation and azimuth would be the only tracking data for the flight, indicating a peak altitude of 1.7 miles (2.7 km), the highest of any of Goddard's rockets. The rocket fell a mile toward the ground before deploying its parachute. By this time, the rocket was rushing downward at about 300 mph (500 km/hr), and the

Test L-13 was the highest-flying of Goddard's 35 liquid propellant rocket flights. (courtesy of Clark University Goddard Library Archives)

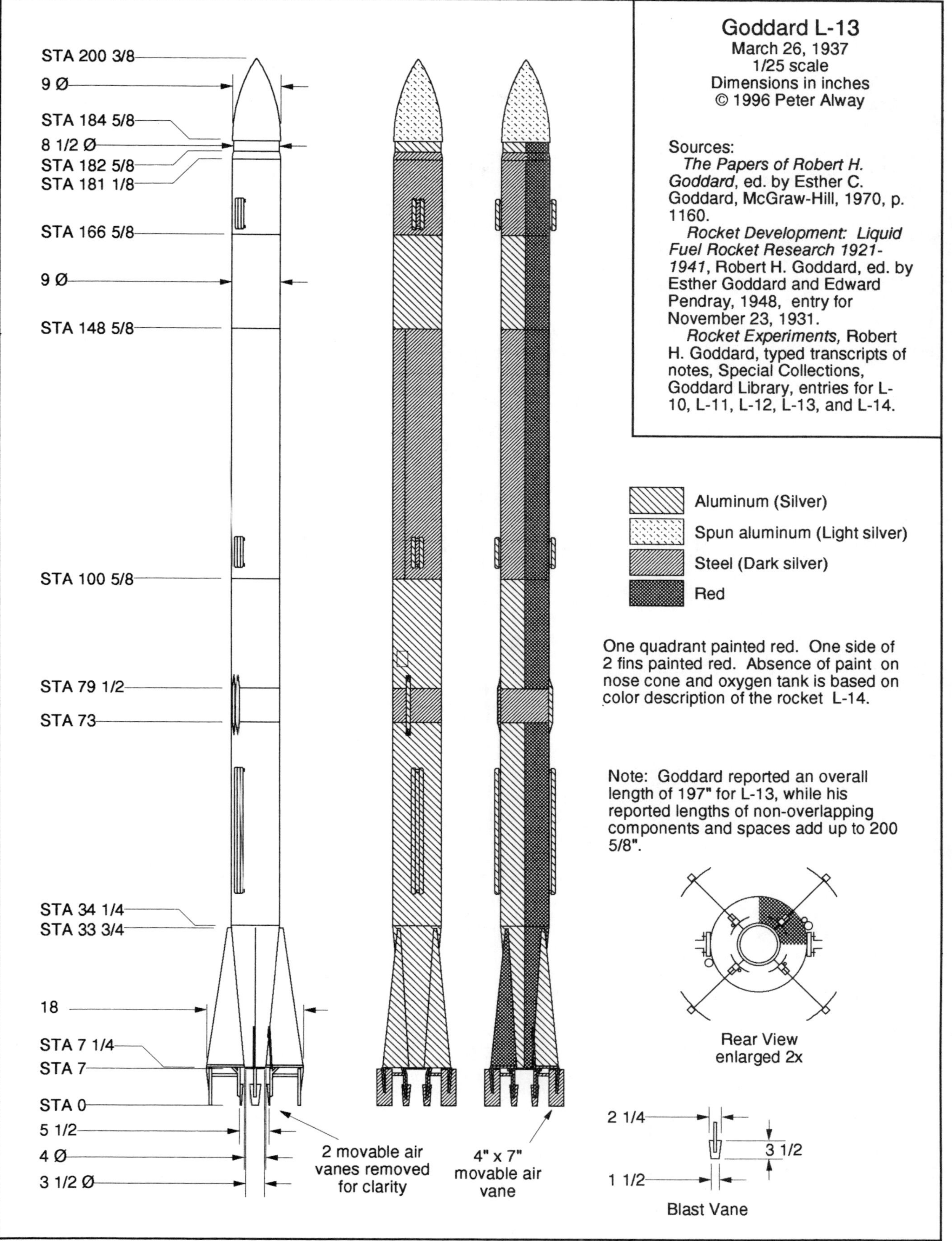
Goddard L-13
March 26, 1937
1/25 scale
Dimensions in inches
© 1996 Peter Alway
Sources:
The Papers of Robert H. Goddard, ed. by Esther C. Goddard, McGraw-Hill, 1970, p. 1160.
Rocket Development: Liquid Fuel Rocket Research 1921-1941, Robert H. Goddard, ed. by Esther Goddard and Edward Pendray, 1948, entry for November 23, 1931.
Rocket Experiments, Robert H. Goddard, typed transcripts of notes, Special Collections, Goddard Library, entries for L-10, L-11, L-12, L-13, and L-14.
STA 200 3/8
9 Ø
STA 184 5/8
8 1/2 Ø
STA 182 5/8
STA 181 1/8
STA 166 5/8
9 Ø
STA 148 5/8
STA 100 5/8
STA 79 1/2
STA 73
STA 34 1/4
STA 33 3/4
18
STA 7 1/4
STA 7
STA 0
5 1/2
4 Ø
3 1/2 Ø
2 movable air vanes removed for clarity
4" x 7" movable air vane
Aluminum (Silver)
Spun aluminum (Light silver)
Steel (Dark silver)
Red
One quadrant painted red. One side of 2 fins painted red. Absence of paint on nose cone and oxygen tank is based on color description of the rocket L-14.
Note: Goddard reported an overall length of 197" for L-13, while his reported lengths of non-overlapping components and spaces add up to 200 5/8".
Rear View enlarged 2x
2 1/4
3 1/2
1 1/2
Blast Vane

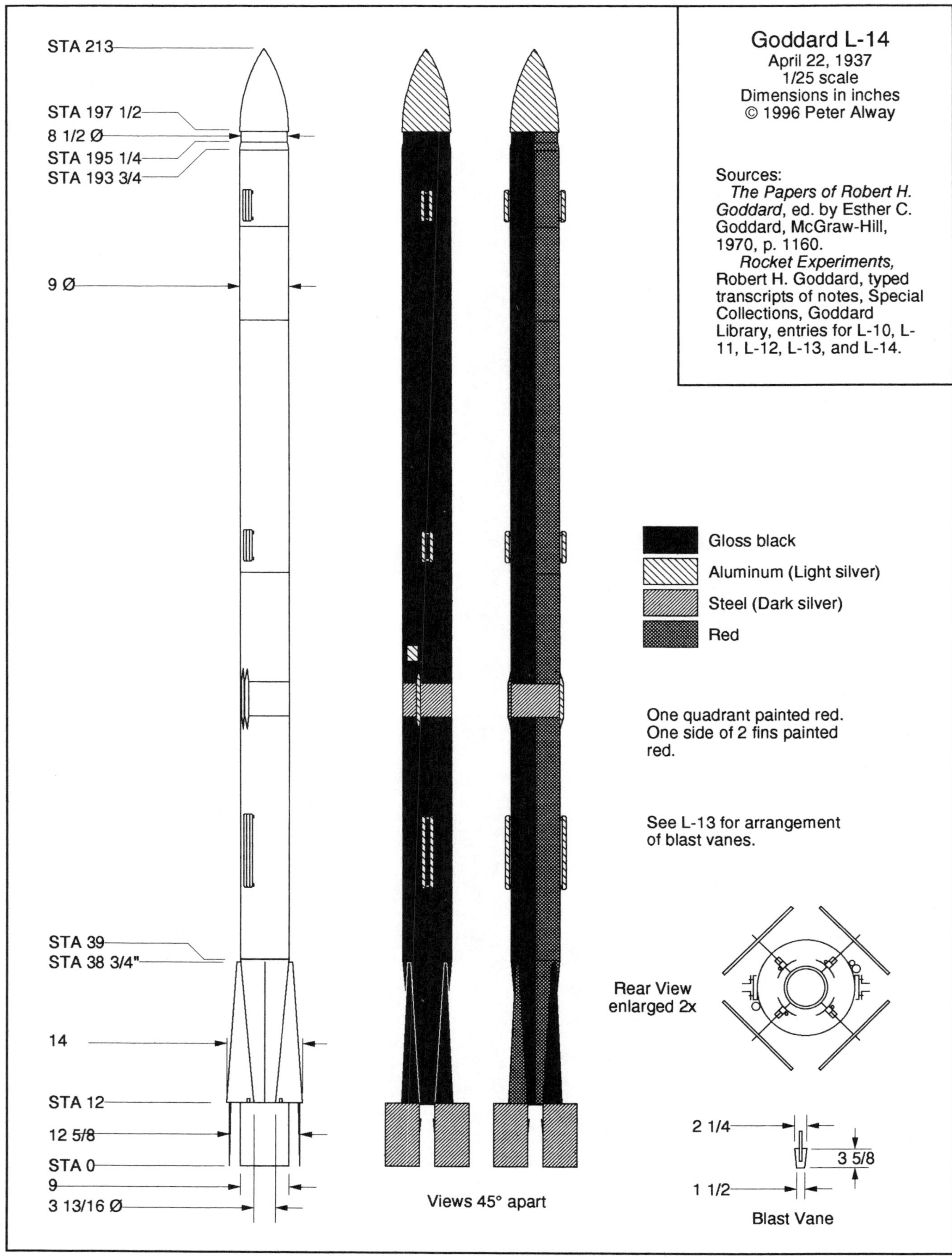
STA 213
STA 197 1/2
8 1/2 Ø
STA 195 1/4
STA 193 3/4
9 Ø
STA 39
STA 38 3/4"
14
STA 12
12 5/8
STA 0
9
3 13/16 Ø
Views 45° apart
Goddard L-14
April 22, 1937
1/25 scale
Dimensions in inches
© 1996 Peter Alway
Sources:
The Papers of Robert H. Goddard, ed. by Esther C. Goddard, McGraw-Hill, 1970, p. 1160.
Rocket Experiments, Robert H. Goddard, typed transcripts of notes, Special Collections, Goddard Library, entries for L-10, L-11, L-12, L-13, and L-14.
Gloss black
Aluminum (Light silver)
Steel (Dark silver)
Red
One quadrant painted red.
One side of 2 fins painted red.
See L-13 for arrangement of blast vanes.
Rear View enlarged 2x
2 1/4
3 5/8
1 1/2
Blast Vane

parachute ripped from its lines.

Rocket L-14 was designed to correct the deficiencies of L-13. It bore gloss black enamel over three fourths of its surface to improve contrast against the dusty New Mexico sky (Goddard hypothesized that the gloss paint would produce less air resistance than a matte finish). Its smaller fixed fins would reduce weathercocking, and huge movable air vanes would insure a vertical trajectory after engine burnout. The flight of L-14 did not live up to expectations, tilting out of the tower and crashing a mile away.

Rocket L-15 carried retractable air vanes—aerodynamic control surfaces held flat against the rocket. When needed to correct the rocket's flight, they would flap outward into the slipstream of air rushing past the rocket (photo on page 2). Another innovation was a wire-wound pressure tank. Goddard had found that the tensile strength-to-weight ratio of wire exceeded that of sheet metal. By making the pressure tank wall extra-thin, and winding it with wire, he was able to save precious weight without sacrificing strength. On May 19, 1937, rocket L-15 made a beautiful vertical ascent, followed by a long slow descent on its parachute. Goddard was pleased with the retractable air vanes; he would use them on all but two of his remaining flights, L-16 and L-17.

L-13 Specifications

Empty weight	100 lb
Duration	22.3 sec
Altitude	9000 ft (2700 m)
Length	16 ft 5 in (5.00 m)
Diameter	9 in (229 mm)

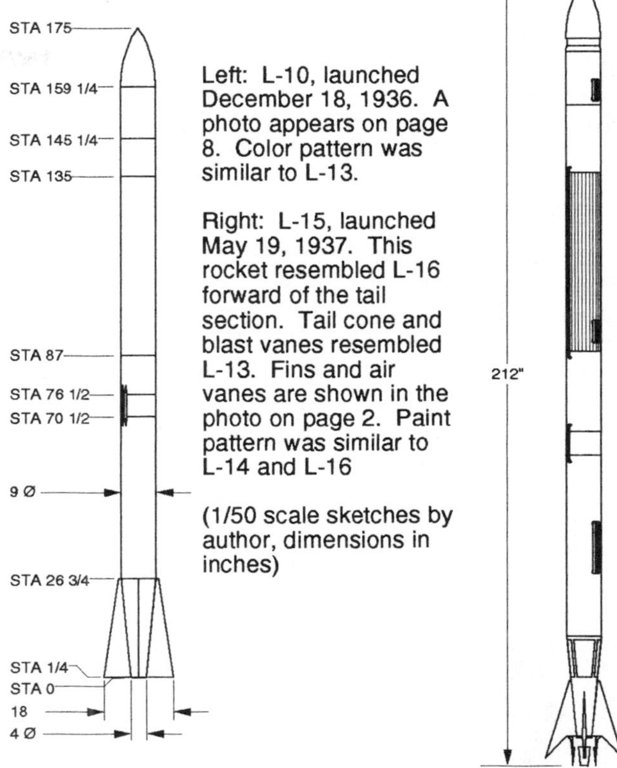

Left: L-10, launched December 18, 1936. A photo appears on page 8. Color pattern was similar to L-13.

Right: L-15, launched May 19, 1937. This rocket resembled L-16 forward of the tail section. Tail cone and blast vanes resembled L-13. Fins and air vanes are shown in the photo on page 2. Paint pattern was similar to L-14 and L-16

(1/50 scale sketches by author, dimensions in inches)

Rocket L-14's large square steering vanes were intended to insure vertical flight after burnout. Black paint was to improve visibility against dusty skies. Red quadrant is just visible along top. (courtesy of Clark University Goddard Library Archives)

L-16 and L-17: The Movable Casing Rocket

Rocket L-16 was to try a new sort of control system. While simple and effective, the blast vanes of previous rockets significantly reduced thrust. L-16 would be a test of a "movable casing rocket," whose gimballed engine would swivel up to 2.5° without loss of thrust. The whole tail section moved with the engine, so the fins doubled as control surfaces.

On July 28, 1937, L-16 flew from the Roswell site. When the rocket deviated from vertical by more than 5°, its gyroscope triggered a pneumatic bellows to swing the tail section to the correct side to straighten the trajectory. The rocket reached an altitude of 2055 feet (616 m), and then began falling sideways. A barometric switch deployed its parachute, allowing recovery for re-flight.

For the second flight, test L-17, Goddard rigged up a bungee-spring catapult to give the rocket a boost off the pad. On August 26, the rocket rose again as L-17. The launch was assisted by a catapult, giving the rocket an apogee of 2000 feet (600 m). But when the parachute deployed, the rocket broke apart, one half floating gently on its parachute and the other half careening into the desert floor. But after counting seven in-flight course corrections, Goddard was satisfied that the movable casing design would eventually become part of his ultimate sounding rocket. The mechanism was complex, however, so he set it aside for his remaining flight tests, and returned to the system of vanes tested on L-15.

L-16 Specifications

Empty weight	93.3 lb (42 kg)
Thrust	477 lb (2120 N)
Altitude	2055 ft (616 m)
Length	18 ft 5 1/2 in (5.61 m)
Diameter	9 in (229 mm)

Al Kisk, Esther Goddard's brother, demonstrates the gimballing of rocket L-16's tail section. While Robert Goddard built only one movable tail section, he felt that this would be the ultimate control mechanism for his final sounding rocket. (courtesy of Clark University Goddard Library Archives)

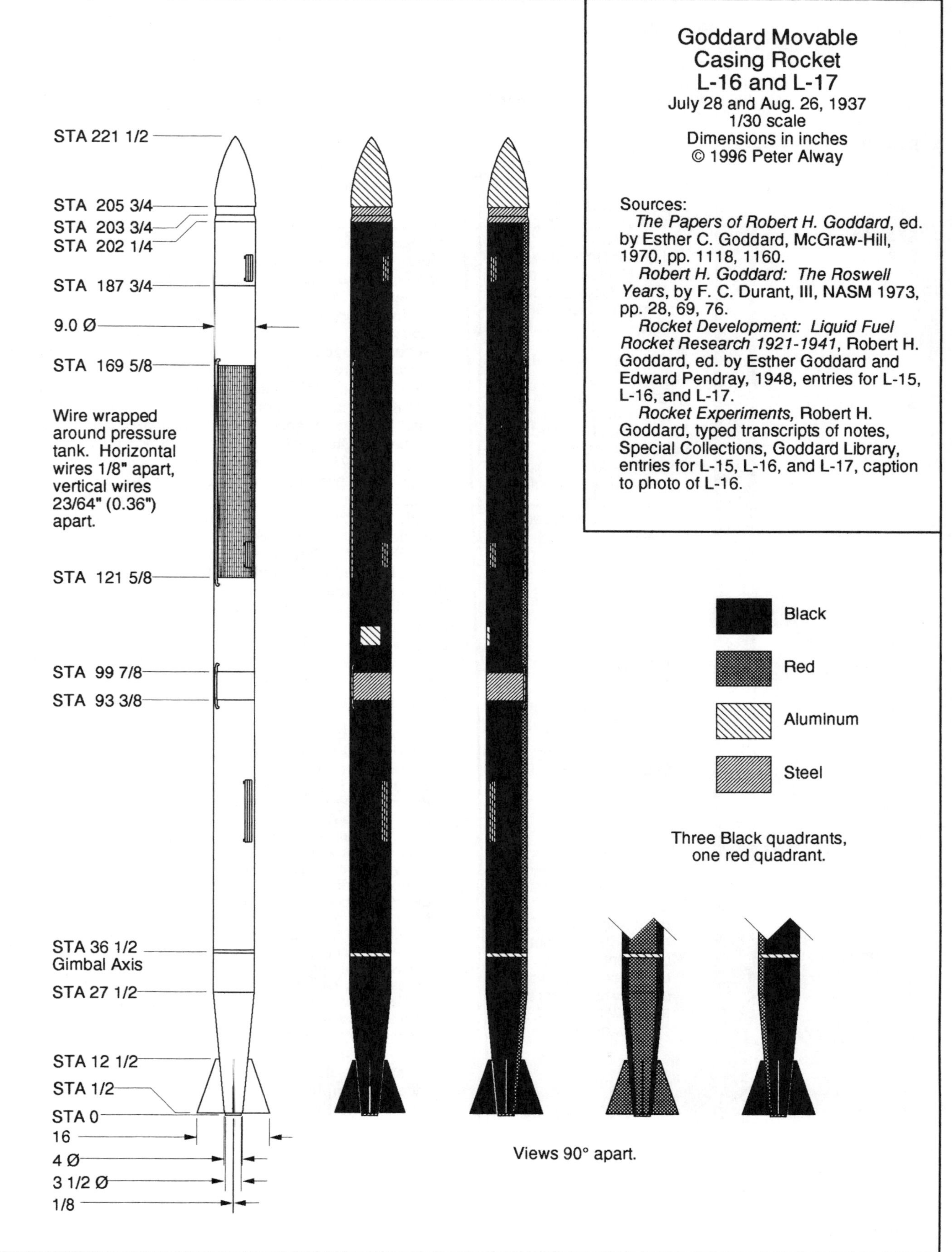

Views 90° apart.

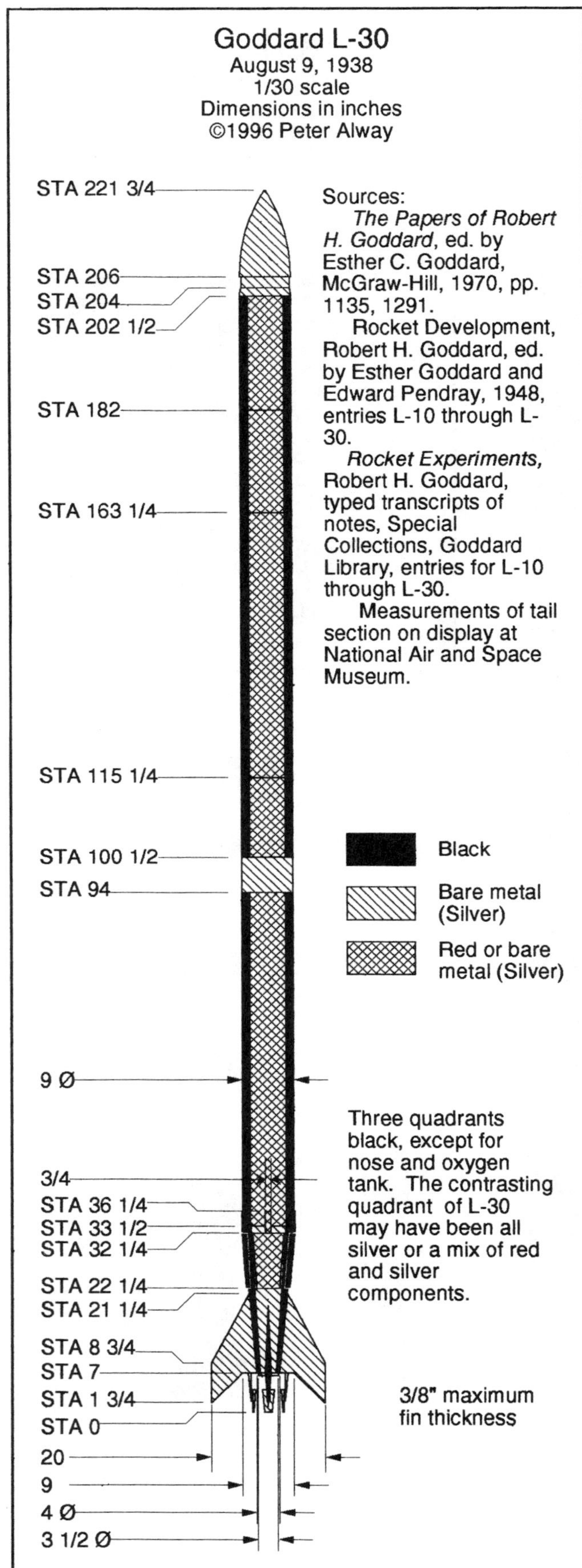

The Last of the L-Series Rockets

With the remaining L-series rockets, Goddard altered his design in hopes of attaining altitudes that would justify further work. While the professor was pleased with the movable casing design, the mechanism was too complex for a test rocket. He would save that feature for the ultimate sounding rocket.

Again, Goddard turned to a bellows pump to pressurize the propellant tanks with vapor from a liquid nitrogen tank. After a series of static tests, rocket L-21 lifted off on November 24, 1937. The rocket never developed full thrust, and crashed just a hundred feet from the tower.

Goddard took a different approach for the following flights, slightly increasing the pressure provided by a gaseous nitrogen tank. Flight L-26, on March 6, 1938, suffered a premature cutoff of oxygen. L-27 performed better, reaching 2170 feet on March 17. For the next flight, Goddard invited a committee of the NAA to verify his rocket's performance with an officially calibrated barometer. On April 20, rocket L-28 ascended vertically to a height of 4215 ft. As the rocket descended, the observers begged the parachute to open, but to no avail. The barograph was flattened into a half-inch-thick lump of scrap metal, and the gasoline remaining in the tank had washed the ink from the recorded chart. Rocket L-29, launched on May 28, crashed before reaching 150 feet.

Dr. Goddard's luck would change with test L-30. For the worse. Goddard's crew loaded the rocket into the tower for a launch on June 14, 1938. But high winds forced a postponement until June 15. On the 15th, a tornado destroyed the tower, with the rocket inside.

Goddard and his stoic crew rebuilt the tower and repaired the rocket. On August 9, rocket L-30 lifted off. Dr. Marjorie Alden of the NAA committee described the flight:

> At about 6:30, Dr. Goddard threw the switch; there was a metallic click followed in a moment by an explosion, and then a steady roar. The flame, first yellow, became white, and in about three seconds after the contact, Dr. Goddard released the anchoring weights and the rocket began to rise. Its rise from the tower was slow, and after clearing tower, it turned into the wind. Its great acceleration was easily apparent, and after a moment its direction corrected to vertical, altered into the wind, and corrected at least once again, and then remained almost vertical. The rocket continued to move with increasing speed, and left a slight trail of bluish white smoke. Its size diminished so rapidly that by the time the fuel was exhausted it was almost invisible. It continued to rise almost vertically, then turned horizontally, and the cap came off releasing the parachute. The flight, with the parachute release, was the most thrilling sight I have ever witnessed. While the barograph recorded 3294 feet, Goddard's ground trackers measured 4920 feet. The NAA witnesses were persuaded by Goddard's calculations.

The tail section of this rocket is on display at the National Air and Space Museum. This was the last of Goddard's simplified test designs. He turned directly to creating a complete, pump-fed rocket.

L-30 Specifications

Launch Weight	146 lb (66 kg)
Empty weight	90.6 lb (41 kg)
Length	18 ft 5 3/4 in (5.63 m)
Diameter	9 in (229 mm)

P-23 and P-31: Pump Fed Rockets

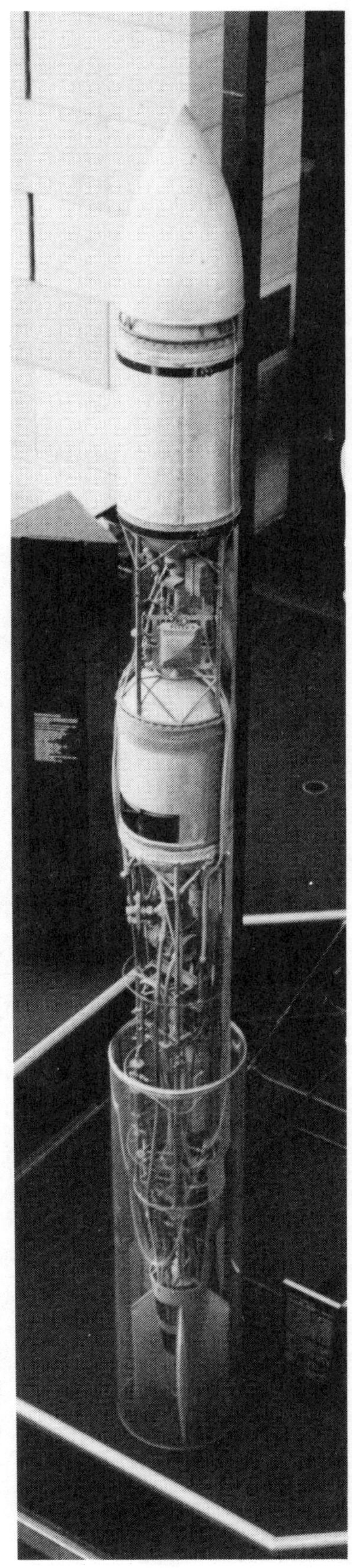

Each of Goddard's rockets to date represented an incremental advance in their inventor's knowledge. But none was really complete. Goddard envisioned his ultimate rocket with a long list of components: liquid fuels in lightweight, low-pressure tanks; turbine-powered centrifugal pumps to force propellants into the combustion chamber; a high-pressure, high-efficiency combustion chamber; a gyroscopic guidance system directing the rocket; blast vanes vectoring thrust during the powered phase of flight; and movable air vanes operating after burnout. Upon the completion of the L-series tests, only one element of the rocket remained unproven—a chemically powered turbine driving a pump to force liquid oxygen and gasoline into a combustion chamber.

Before embarking on this final project, Robert and Esther Goddard took a well earned vacation in France, and on returning through New York, met with Harry Guggenheim. Guggenheim arranged meetings with representatives of the Massachusetts Institute of Technology, the National Advisory Committee on Aeronautics, and the California Institute of Technology. All urged Goddard to collaborate. Caltech's Theodore Von Kármán offered a partnership with Goddard. Goddard refused, leaving the institute to develop liquid-fueled rockets on its own.

Goddard dusted off the turbopumps he tested back at Clark in 1934, and built several new designs. He dedicated the rest of 1938 to testing the turbine-driven centrifugal pumps. The high pressure provided by the pumps allowed a standard 5 3/4" (146 mm) combustion chamber to produce over twice the thrust it could generate under tank pressures. It took Goddard almost all of 1939 to incorporate the pumps into a complete rocket for a static test. The professor had first conducted a series of tests with 9" (229 mm) tanks before settling on an 18"-diameter (457 mm), 22 ft tall (6.7 m) P-type flight model. A gas generator, burning gasoline and liquid oxygen, powered the turbines which, in turn, powered the gasoline and oxygen pumps that fed the rocket engine. The tanks were pressurized with nitrogen gas evaporated from a liquid supply. Finally, on November 18, 1939, a complete rocket, P-13, was ready for a static test.

Goddard did not know of the year's developments in Germany's secret missile program. The A-5, created under the supervision of Wernher von Braun, had already flown to an altitude of 7 miles, in a rocket sharing most of its major components with Goddard's static test model. As the 1930's ended, the German missile program was well underway.

By February 9, 1940, Goddard was ready to launch. But rocket P-15 failed to leave the ground—the oxygen pump clogged with ice and popped open, blowing off part of the rocket casing. Flights P-16, P-21, and P-22 failed to leave the ground as well. In his last years at Roswell, Goddard faced a pattern of failures and incremental progress. Launch attempts would fail, a series of static tests would follow until a problem appeared to be solved, and then a flight test would fail.

On August 9, 1940, Robert Goddard pressed a telegraph key to begin the ignition sequence on flight test P-23. The rocket exploded to life with a pop that Goddard had come to expect with his pump rockets. A white indicator light indicated the rocket was producing thrust. Then a red light, indicating thrust sufficient for liftoff turned on. Goddard pressed a release key, and the rocket crept up its tower atop a yellow flame. Goddard estimated a speed of just 10-15 mph as the rocket cleared the tower. In spite of calm conditions, the rocket tipped off the top of the tower, and the steering vanes lacked the power to right the rocket. It tipped over as it reached about 200-300 feet, and careened into the ground under power. The crash tore open the propellant tanks, and the observers saw, and then felt, a tremendous explosion.

More static tests. More modifications to pumps, and engines. More static tests.

Goddard's last rocket, a pump-fed rocket now on display at the National Air and Space Museum. Color is not representative of the two rockets that took flight. (photo by author)

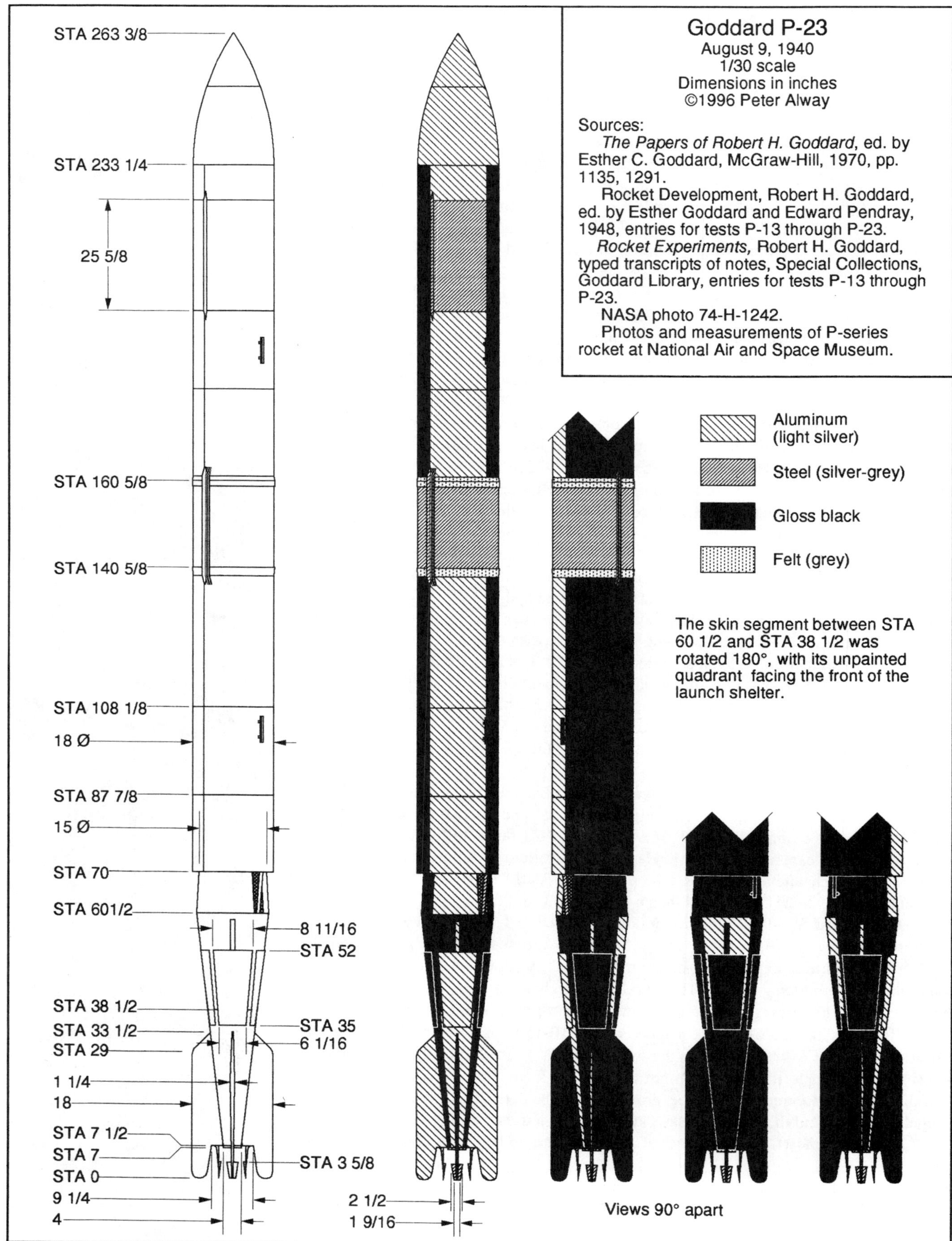
STA 263 3/8
STA 233 1/4
25 5/8
STA 160 5/8
STA 140 5/8
STA 108 1/8
18 Ø
STA 87 7/8
15 Ø
STA 70
STA 601/2
8 11/16
STA 52
STA 38 1/2
STA 33 1/2
STA 29
STA 35
6 1/16
1 1/4
18
STA 7 1/2
STA 7
STA 0
STA 3 5/8
9 1/4
4
2 1/2
1 9/16
Goddard P-23
August 9, 1940
1/30 scale
Dimensions in inches
©1996 Peter Alway
Sources:
The Papers of Robert H. Goddard, ed. by Esther C. Goddard, McGraw-Hill, 1970, pp. 1135, 1291.
Rocket Development, Robert H. Goddard, ed. by Esther Goddard and Edward Pendray, 1948, entries for tests P-13 through P-23.
Rocket Experiments, Robert H. Goddard, typed transcripts of notes, Special Collections, Goddard Library, entries for tests P-13 through P-23.
NASA photo 74-H-1242.
Photos and measurements of P-series rocket at National Air and Space Museum.
Aluminum (light silver)
Steel (silver-grey)
Gloss black
Felt (grey)
The skin segment between STA 60 1/2 and STA 38 1/2 was rotated 180°, with its unpainted quadrant facing the front of the launch shelter.
Views 90° apart

Goddard's first pump-fed rocket before a launch attempt on August 1, 1940. This rocket flew a week later in Test P-23. Note that one skin section has been removed at this stage of preparation. (NASA photo 74-H-1242)

It was clear that the pump-fed rocket was a bigger challenge than Goddard had thought. Perhaps the P-series pumps were simply beyond the ability of Goddard's crew to fabricate, although this thought seems not to have occurred to the professor. Goddard would not assume success, either. He would equip the next rocket with an emergency system to shut down the engine and deploy a parachute if it pointed below the horizontal.

On May 8, 1941, rocket P-31 was ready to fly. Goddard pressed the key to start the ignition sequence. The ground equipment began to work, but its workings were invisible from the launch shelter. Something was wrong—there was no pop, no roar—no sound. The crew waited. A minute ticked by. No ignition. Another minute. Still nothing. The rocket was still live, loaded with gasoline and liquid oxygen, valves open. Another minute. And another. Then, five minutes after the launch command, there was the roar of a flame. All eyes turned to the rocket. A white light indicated lift. A red light indicated enough thrust for a flight. The umbilical came out of the rocket, and Goddard pressed the release button. This time the rocket rose faster, reaching two or three times the height of the 80-foot tower before tipping over. As the nose dipped 20° toward the ground, the flame dimmed, and the nose came off, releasing its parachute too close to the ground to deploy fully. The rocket, nearly full of fuel, burst into flames on impact. For the last time, a spaceship had crashed into the desert outside Roswell, New Mexico. But thanks to the new emergency system, the rocket was largely salvageable.

Again, Goddard embarked on a series of static tests and launch attempts. On October 10, 1941, rocket P-36 came to life. The lift lights flashed quickly, and Robert Hutchings Goddard pressed the release key one last time. But the rocket jammed in the tower. World War II ended the research program. Goddard's last rocket now stands on display near the entrance to the National Air and Space Museum.

P-23 Specifications

Launch weight	236 lb (107 kg)
Empty weight	161 lb (73 kg)
Thrust	700 lb (3000 N)
Altitude	300 ft (100 m)
Length	21 ft 11 5/8 in (6.69 m)
Diameter	18 in (457 mm)

Goddard felt it was his duty to apply himself to the war effort. The Germans were within a year of launching the V-2 missile, and Goddard had essentially invented its major systems on his own, with or without the knowledge of the German researchers. But there was no interest in ballistic missiles in the US military. The professor applied his genius to a more mundane problem—rocket assisted take-off for Navy seaplanes. In a 1942 test, Goddard's finicky rocket succeeded—first in getting an amphibious PBY Catalina airborne, and second in setting fire to it. The Navy turned to solid propellants, and assigned Goddard to trivial tasks as a "consultant."

As news began to filter out from Europe about the V-2, Goddard suffered pangs of recognition. In a report to Harry Guggenheim, Goddard laid out a table comparing the features of the V-2 and his own P-series rocket. The descriptions varied only in the choice of chemicals for fuel and turbine drive for the pumps. While he suspected that the Germans had stolen his ideas, Goddard began to recognize that his reluctance to share his work with his American colleagues had served no one but the Germans. Mostly he regretted more that his own government had failed to see the value of his work.

In the spring of 1945, the professor and his crew had the opportunity to inspect a V-2 captured in the final days of the war in Europe. Henry Sachs, who had ignited the world's first liquid propelled rocket 19 years earlier, commented "It looks like ours, Dr. Goddard."

"Yes, Mr. Sachs" he replied in a hoarse whisper, "it seems so." Goddard's whisper was a symptom of advancing throat cancer. On August 10, 1945, a day after reading of the second atomic bomb blast in Japan, Robert H. Goddard died.

Goddard wanted to invent the space rocket alone. To a remarkable extent, he succeeded. By the time of his death, Goddard had invented nearly every component of a functional space launch booster. But so had Wernher von Braun's V-2 group in Germany. Frank Malina and his JPL group would fly their independently created Wac Corporal sounding rocket within a month. Malina had, in fact, wanted to collaborate with Goddard, but Goddard refused, keeping his latest rocket literally under wraps when the upstart graduate student visited his Roswell shop. Goddard refused to believe that the age of the lone inventor was over, and that the space rocket would ultimately be built by big teams funded by big governments.

Goddard's achievements are unquestionable: The first use of a de Laval Nozzle in a rocket, the first demonstration of a rocket in a vacuum, the invention of the Bazooka, the first liquid-propelled rocket flight, the first rocket controlled by an inertial guidance system, the first use of thrust vector control in a rocket, the first use of a gimballed engine in a rocket, the first turbopumps built for a liquid propellant rocket, the first pulse-jet engine, and the first liquid propellant rocket cluster are among his accomplishments.

Yet his *contributions* to the conquest of space are open to question. By denying details of his work to colleagues, he forced them to re-invent the modern rocket on their own. For all the similarities between the V-2 and Goddard's rockets, the Germans made little use of Goddard's work. There are conflicting reports about German access to Goddard's patents, and Goddard's brief 1936 Smithsonian report is hardly a manual for missile design—the guidance system is depicted only as a small box in the professor's hand. The report did, however, indicate the direction of Goddard's work; it gave the German's the mistaken impression that the United States was building a ballistic missile, lending credibility to the V-2 program.

Goddard's secrecy, in part inspired by fear of German missile makers, probably contributed to the German superiority in rocketry during WW II. In the US, Goddard's designs did have a strong influence in the creation of Curtiss-Wright's XLR-25-CW-1 which powered the X-2 supersonic aircraft.

For all that Goddard valued his independence, he recognized the magnitude of the challenge of the spaceship. He willingly left the creation of the Moon rocket to future generations. Robert knew he would not live to see explorers travel by rocket to the Moon. He could not have guessed that Esther would.

Robert Goddard left over 200 patents, many filed by Esther after his death. Most were so technical as to conceal their applications to rocketry, yet some may have been of use to the Germans developing the V-2 (US Patent Office lore has it that the Germans had a standing order for all of Goddard's patents). After the war, it was clear that the US Army, Navy, Air Force, and NASA could not engage in rocket work without infringing on the patents. A million-dollar settlement guaranteed Esther Goddard a comfortable retirement, secure in the knowledge that wherever rocket science went, Robert Goddard had been there first.

Goddard's Liquid-Propelled Rocket Launches

Date	Name	Altitude (Range)	Configuaration
March 16, 1926	(1st liquid)	41'	open frame tractor, ~120" long
April 3, 1926			open frame tractor
December 26, 1928	Hoopskirt	(200')	open frame tractor, 176" long
July 17, 1929	(reported as plane crash)	90'	open frame, 138" long
December 30, 1930	Test No. 72	2000'	open frame, 132" long
September 29, 1931	Test No. 73	180'	9 3/8" x 119"
October 13, 1931	Test No. 74	1700'	9 3/8" x 93"
October 27, 1931	Test No. 75	1330'	9" x 103"
April 19, 1923	Test No. 77 (1st Gyro rocket)	135'	12" x 129 1/2"
February 16, 1935	A-3	220'	9" x 162 3/8"
March 8, 1935	A-4	1000'	9" x 175" (see A-5)
March 28, 1935	A-5	4800'	9" x 177 3/4"
May 31, 1935	A-8	(5500')	9" x 181 1/2" (see A-5)
June 25, 1935	A-9	120'	9" x 183 1/2" (see A-11)
July 12, 1935	A-10	6600'	9" x 183 1/2" (see A-11)
October 29, 1935	A-14	4000'	9" x 183 1/2" (see A-11)
July 31, 1936	L-5	200'	18" x 154"
October 3, 1936	L-6	200'	18" x 160 1/2"
November 7, 1936	L-7 (4-chamber rocket)	200'	18" x 162 1/2"
December 18, 1936	L-10	(2000')	9" x 175 1/2"
February 1, 1937	L-11	1870'	9" x 199 1/2" (see L-13)
February 27, 1937	L-12	1500'	9" x 209 1/4" (see L-13)
March 26, 1937	L-13	9000'	9" x 197"
April 22, 1937	L-14	(5000')	9" x 213"
May 19, 1937	L-15	3250'	9" x 212"
July 28, 1937	L-16 (movable casing)	2200'	9" x 221"
August 26, 1937	L-17 (movable casing)	2000'	9" x 221"
November 24, 1937	L-21	(100')	9" x 208" (see L-30)
March 6, 1938	L-26	over 500'	9" x 210" (see L-30)
March 17, 1938	L-27	2170'	9" x 216" (see L-30)
April 20, 1938	L-28 (NAA barograph)	4215'	9" x 221 3/4" (see L-30)
May 26, 1938	L-29 (NAA barograph)	150'	9" x 221 3/4" (see L-30)
August 9, 1938	L-30 (NAA barograph)	4290'	9" x 221 3/4"
August 9, 1940	P-23 (pump-fed rocket)	300'	18" x 263 3/8"
May 8, 1941	P-31 (pump-fed rocket)	250'	18" x 263 3/8" (see P-23)

The American Rocket Society

On April 4, 1930, a dozen space enthusiasts—mostly science fiction writers from publisher Hugo Gernsback's magazine organization—gathered in a New York apartment to create the American Interplanetary Society, modeled after the interplanetary societies of Europe. After the organizational meeting, the group decided to give their club a public launch with a speech by the noted French rocket experimenter and author of *L'Astronautique,* Robert Esnault-Peltier. The speech would accompany a showing of *A Girl in the Moon,* an English translation of Fritz Lang's silent *Frau im Mond.* In their nerdish enthusiasm, the members in charge of this entertainment hacked out the romantic subplots from the film, leaving the technical scenes that had been prepared with the advice of rocket pioneer Hermann Oberth. When Esnault-Peltier withdrew from the program due to illness, the Society's vice president, Edward Pendray, stepped in to read Esnault-Peltier's address on the future of space rocketry. The society grew quickly, attracting members from around the country, and a few foreign members.

The society's favorite member, however, refused to participate actively in the group. Robert Goddard joined simply to receive its magazine, *Astronautics,* to keep up with other researchers. For years, the society's leaders didn't even realize that Goddard had successfully flown a liquid propellant rocket.

When Pendray had the opportunity to vacation with his wife in Europe in 1931, he took advantage of the trip to meet with Germany's Verein für Raumschiffahrt (Society for Space Travel, abbreviated VfR). Pendray was warmly welcomed by the group, and was even shown a static test of a VfR rocket. With the help of VfR secretary Willy Ley, he made detailed sketches of the VfR's Mirak rockets and presented them to the AIS.

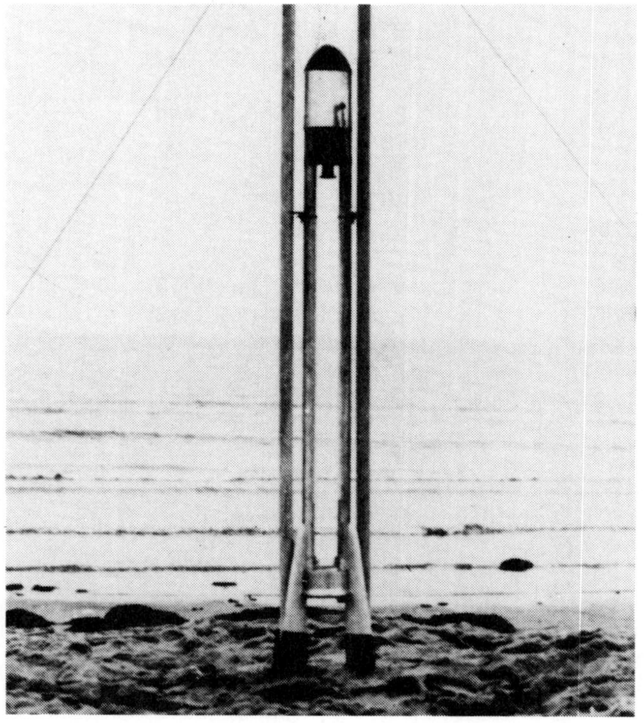

The American Rocket Society's Rocket No. 2. (Smithsonian Institution negative 87-17038)

ARS No. 2

Pendray encouraged the Interplanetary Society to duplicate the German efforts (this would eventually lead to a change in name to the American Rocket Society in 1934; strangely, the rockets the society built before the name change are universally referred to as ARS designs). In November of 1932, ARS rocket No. 1 emerged from a member's basement machine shop. The rocket borrowed the VfR's design for the 2-Stick Repulsor, whose parallel oxygen and fuel tanks trailed behind the engine. The engine, designed by Pendray and H. F. Pierce, burned gasoline with liquid oxygen. The combustion chamber took the form of a cylinder with hemispherical ends, one of which bore a long, conical nozzle. A static test on a New Jersey farm damaged the fragile machine.

Bernard Smith was a 22-year-old high school dropout who worked odd jobs and hung out around New York's libraries and museums. When Smith attended a Society meeting and suggested improvements to the rocket, Pendray handed him the wreckage and set him to work creating ARS No. 2. The depression had hit Smith particularly hard, so his motivation for rocket construction was simple. "It was a lousy planet; the rocket ship was the only way to get off it," he later recalled.

Smith trimmed the rocket down into a flightworthy machine. He simplified valves, removed a water cooling system, and removed unnecessary metal parts. His additions were components scavenged from a coffee pot, a gas range, and whatever other scrap materials he could find. ARS No. 2 featured a 3" (76 mm) diameter by 6" (152 mm) long engine mounted between two aluminum tubes containing liquid oxygen and gasoline. A small hole in the tip of the nose cone admitted air to cool the engine. The gaseous oxygen boil-off forced liquid oxygen up a tube into the engine, while pressurized nitrogen forced gasoline into the engine. Aluminum paint protected four balsa wood stabilizing fins from the blast of the rocket exhausts. On May 14, 1933, ARS experimenters, accompanied by two newsreel camera crews, carried their rocket to the beach at Marine Park, Great Kills, Staten Island for its first flight.

After a brief static firing, Pendray and Smith refueled the rocket. Three minutes before launch, member Alfred Best lit a fuse—a gasoline-soaked rag—under the nozzle. At the appointed moment of launch, Smith, sheltered in a dugout, pulled a cord leading to the propellant valves. The valve handle fell off. He ran from the dugout and re-attached the handle. Smith pulled the cord on his way back to the shelter, and the rocket lifted off.

Once it cleared the 15-foot (5 m) launch rail, ARS No. 2 turned into the wind, 90° from the planned seaward trajectory. The rocket was expected to reach an altitude of about a mile (1.6 km). But when the rocket reached 250 feet (75 m), the oxygen tank ruptured with a loud pop. The flaming remains splashed into Lower New York Bay, to be recovered by two boys who happened by in a rowboat.

ARS No. 2 Specifications

Launch weight	~15 lb (7 kg)
Thrust	60 lb (270 N)
Duration	2 sec
Total impulse	120 lb-sec (540 N-sec)
NAR designation	I270
Altitude	250 ft (75 m)
Length	7 ft (2.1 m)
Diameter	7 in (18 cm)

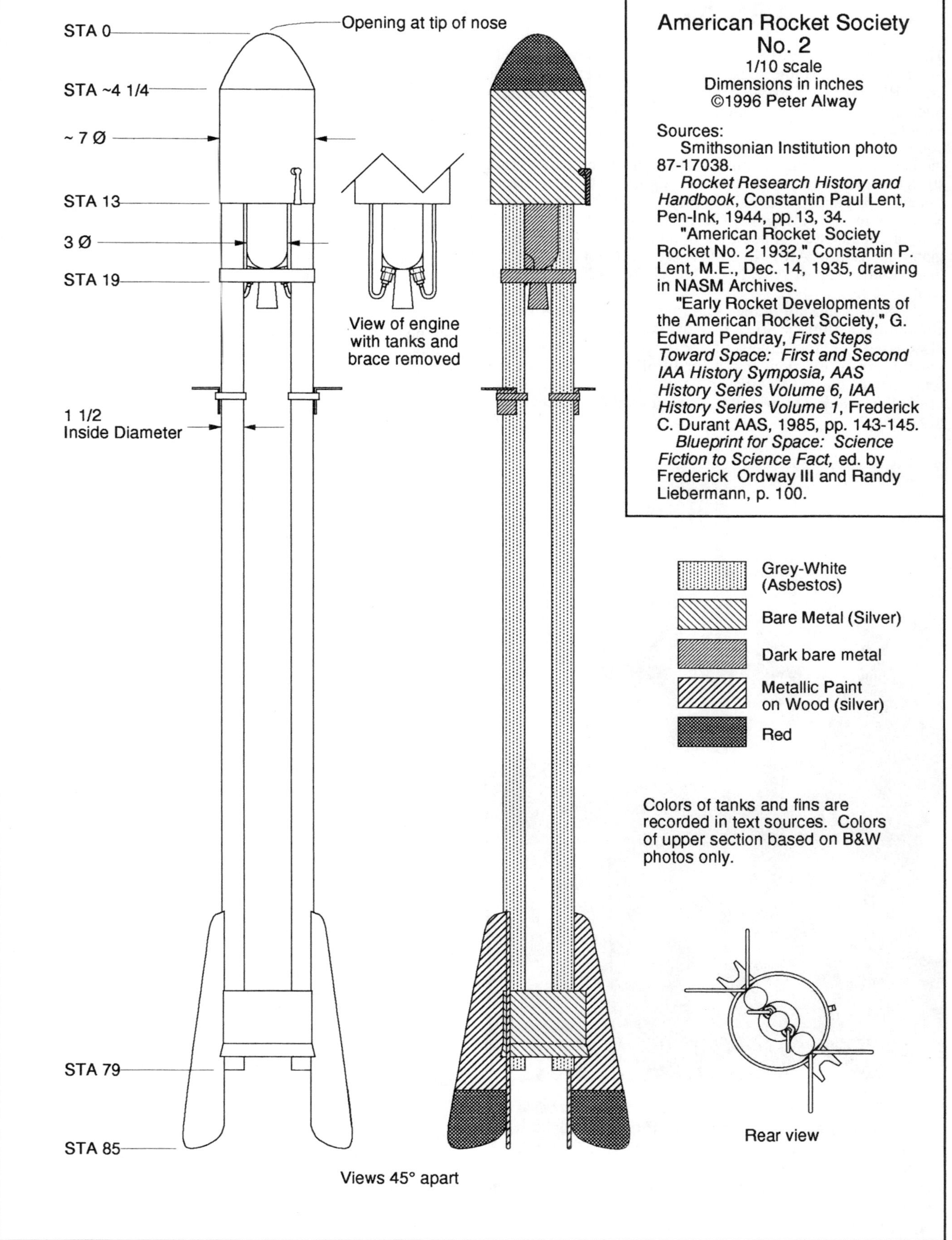

American Rocket Society No. 2

1/10 scale
Dimensions in inches

Sources:

Smithsonian Institution photo 87-17038.

Rocket Research History and Handbook, Constantin Paul Lent, Pen-Ink, 1944, pp.13, 34.

"American Rocket Society Rocket No. 2 1932," Constantin P. Lent, M.E., Dec. 14, 1935, drawing in NASM Archives.

"Early Rocket Developments of the American Rocket Society," G. Edward Pendray, *First Steps Toward Space: First and Second IAA History Symposia, AAS History Series Volume 6, IAA History Series Volume 1*, Frederick C. Durant AAS, 1985, pp. 143-145.

Blueprint for Space: Science Fiction to Science Fact, ed. by Frederick Ordway III and Randy Liebermann, p. 100.

Colors of tanks and fins are recorded in text sources. Colors of upper section based on B&W photos only.

ARS No. 4

The success of ARS No. 2 inspired more Society members to contribute rocket designs. The next to fly was the ARS No. 4. Designed by professional engineer John Shesta, this rocket was modeled after the VfR's 1-Stick Repulsor. Four nozzles protruded from the rear of the combustion chamber, while fuel and oxidizer entered from the center-rear of the chamber. The tanks hung behind in a straight line. In front of the engine was a recovery system—originally to be a helicopter device, but switched to a parachute before flight. The parachute was stowed in a cylindrical canister topped by a blunt cap. Considering the materials used in ARS No. 2, it is probably not coincidental that the nose cap of ARS No. 4 closely resembled a kitchen cookware lid.

Shesta did not bother static testing the ARS-4 before attempting a flight. So it was on the launch pad that he discovered the fuel inlets of the engine were too small, resulting in insufficient thrust for liftoff. When one of the four nozzles burned through, it became clear that the rocket needed a water cooling jacket around the chamber and nozzles.

Edward Pendray (right) readies ARS No. 4 for flight while designer John Shesta (left) watches. Propellant tanks hang below the 4-nozzled combustion chamber, while the cylindrical parachute compartment is above. A water jacket was added around the combustion chamber and four small nozzles before it took flight. (Smithsonian Institution negative no. A-4558-C)

The ARS experimenters modified the rocket and tried again on September 9, 1934. This time the rocket leaped from its launcher, again set up at Marine Park, and flew out over the Atlantic. Altitude was carefully tracked to 382 feet. In horizontal flight, it approached the speed of sound. Parts of the rocket were recovered—the tanks were incorporated into a static test stand.

Pendray and Smith never got their ARS No. 3 off the ground—the concentric gasoline and oxygen tanks conducted heat to the liquid oxygen so quickly that it evaporated before it could fill the tank. The society turned to creating improved engines and giving them static tests, often going to great lengths to avoid suspicious police and fire officials. These tests were not without hazards. A bystander at one test was hospitalized with a broken elbow when she was hit by a fragment of an exploding engine.

The most important of these unflown engines was James Wyld's regeneratively cooled engine. The ARS had learned from the VfR that liquid oxygen was an unsuitable coolant, prone to explosions. Goddard's method of regenerative cooling, "curtain cooling," used a film of swirling gasoline inside the combustion chamber to protect the engine from burn-throughs. But this technique proved unreliable and unscalable to larger engines. Wyld's engine routed gasoline between layers of a double-walled combustion chamber. Once refined, this engine could burn indefinitely without damage. Wyld was not the first to design such a rocket, but his was the first such design published in the open literature in 1938.

With the onset of WW II, the club gave up on experimentation and evolved into a professional society, producing a trade journal, and eventually merged with the Institute of Aerospace Sciences to become today's American Institute of Aeronautics and Astronautics.

The ARS program proved to be a valuable training ground for some of the nation's leading rocket engineers. Four members of the pre-war group—Shesta, Wyld, Pierce, and Lovell Lawrence, Jr.—went on to form Reaction Motors, Incorporated to exploit Wyld's regenerative engine. Their first task during World War II was to produce JATO (Jet assisted take-off) rockets for the Navy. After the war, Reaction Motors produced the powerplants for the X-1, X-15, MX-774, and Viking sounding rocket. Later, Reaction Motors became Thiokol's liquid propellant rocket division, which produced engines for the Sparrow and Bullpup missiles. While the American Interplanetary Society never sent a rocket to the Moon, NASA's Surveyor spacecraft touched down on the Moon under the power of a Thiokol-Reaction Motors vernier rocket engine.

Bernard Smith never did finish high school, but he earned a degree in physics and embarked on a fruitful career that cured him of his urgent need to get off the planet. He eventually become head of the Weapons Development Department at the Naval Ordinance Test Station, where he managed or contributed to several missile development programs, including the RAT, ASROC, Sidewinder, and Shrike. Under his ægis, Project Pilot launched two small satellites into orbit aboard air-launched solid rockets in the summer of 1958 (Smith described the attempts in 1983, but their success was not revealed until the 1990's).

ARS No. 4 Specifications

Expected duration	30 sec
Actual duration	12 sec
Altitude	382 ft (116 m)
Length	7 ft 10 in (2.38 m)
Diameter	3 in (76 mm)

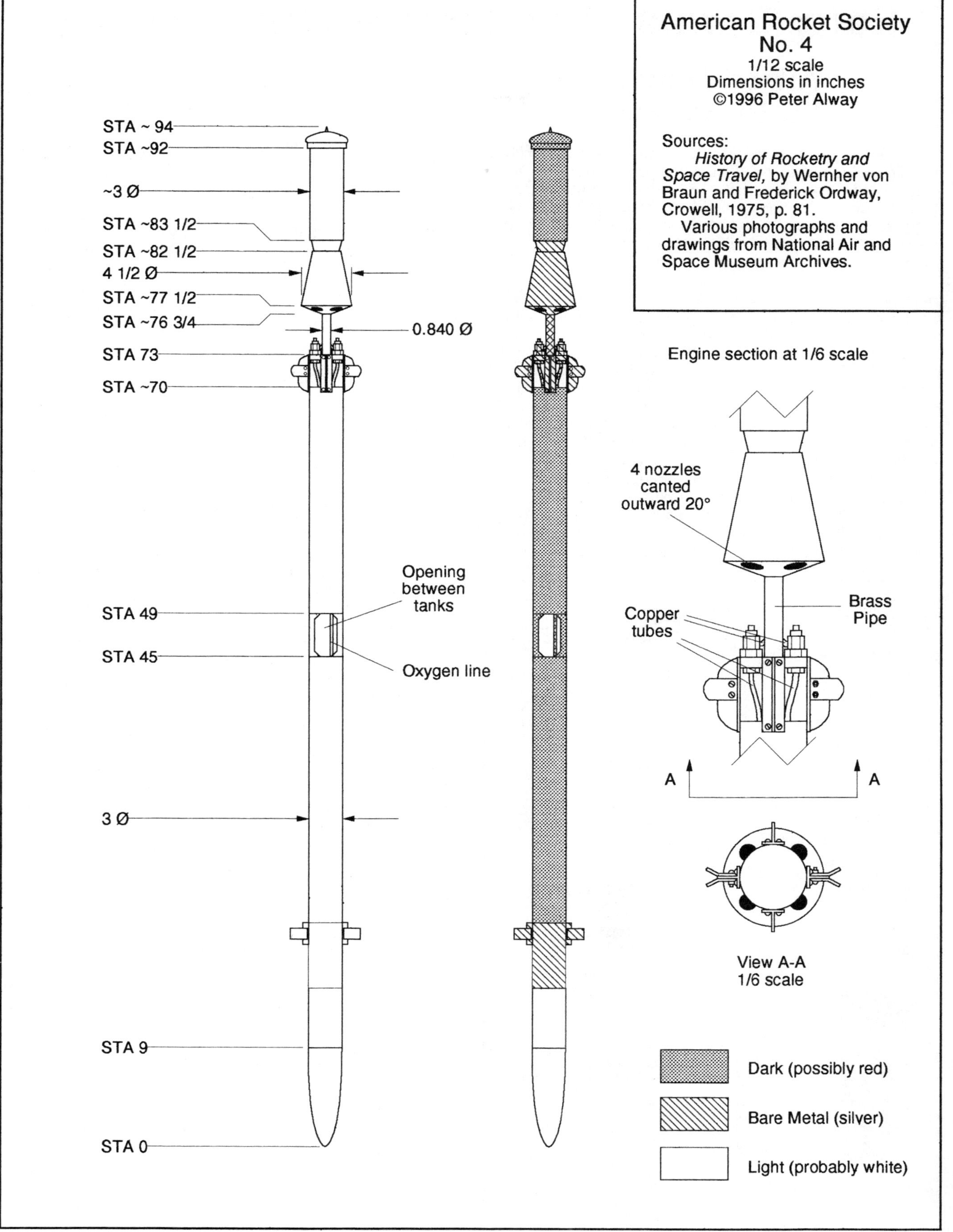
American Rocket Society
No. 4
1/12 scale
Dimensions in inches
©1996 Peter Alway

Sources:
History of Rocketry and Space Travel, by Wernher von Braun and Frederick Ordway, Crowell, 1975, p. 81.
Various photographs and drawings from National Air and Space Museum Archives.

STA ~ 94
STA ~92
~3 Ø
STA ~83 1/2
STA ~82 1/2
4 1/2 Ø
STA ~77 1/2
STA ~76 3/4
0.840 Ø
STA 73
STA ~70
Opening between tanks
STA 49
STA 45
Oxygen line
3 Ø
STA 9
STA 0

Engine section at 1/6 scale
4 nozzles canted outward 20°
Brass Pipe
Copper tubes
A
A
View A-A
1/6 scale

Dark (possibly red)
Bare Metal (silver)
Light (probably white)

Early German Solid Propellant Rocketry

Alfred Maul's Photo Rockets

Before German spaceflight enthusiasts ever launched a liquid fueled rocket, two German experimenters produced solid-fueled rockets that anticipated future space developments. The first was Alfred Maul, a Dresden civil engineer, who patented a photographic rocket in 1903. He was not the first to patent such a device, but his was the first known to see testing and use.

Maul began serious work on a camera rocket system to provide aerial reconnaissance for the Army in 1901. Unlike a balloon, a rocket could be prepared for launch quickly, spent only a few seconds exposed to ground fire, and exposed its operators to relatively little risk. For 11 years, Alfred Maul built progressively larger rockets, and one by one he solved the problems of rocket photography. Fins attached to a long guide stick insured aerodynamic stability, while a gyroscope maintained the pointing of the camera (Maul patented this innovation in 1906). Maul developed a parachute recovery system to insure safe landing of the camera and its photographic plate. He selected commercial rockets for propulsion.

By 1912, under the sponsorship of the Saxon Army, Maul had produced a fully functional system. In eight seconds, the 42 kg (92 lb) rocket reached an altitude of 800 meters (2,600 ft). As the rocket neared apogee, an electro-pneumatic trigger fired the camera. The rocket remained in a vertical orientation to this point; the camera pointed downward within the nose cone to insure a view of the ground. Azimuth pointing was maintained by the gyroscope. The rocket then ejected the nose cone and parachute connected by a cord to the rocket body. A minute later, the rocket touched down. Within a few minutes the camera could be recovered and the plates developed. Some of these photos still exist, showing clear bird's-eye views of villages near the Saxon Army range at Königsbrük.

Maul's invention was battle tested in the Turkish-Bulgarian War of 1912-1913. Anticipating today's reconnaissance satellites, it produced clear photographs of Turkish emplacements for the German-allied Bulgarian Army.

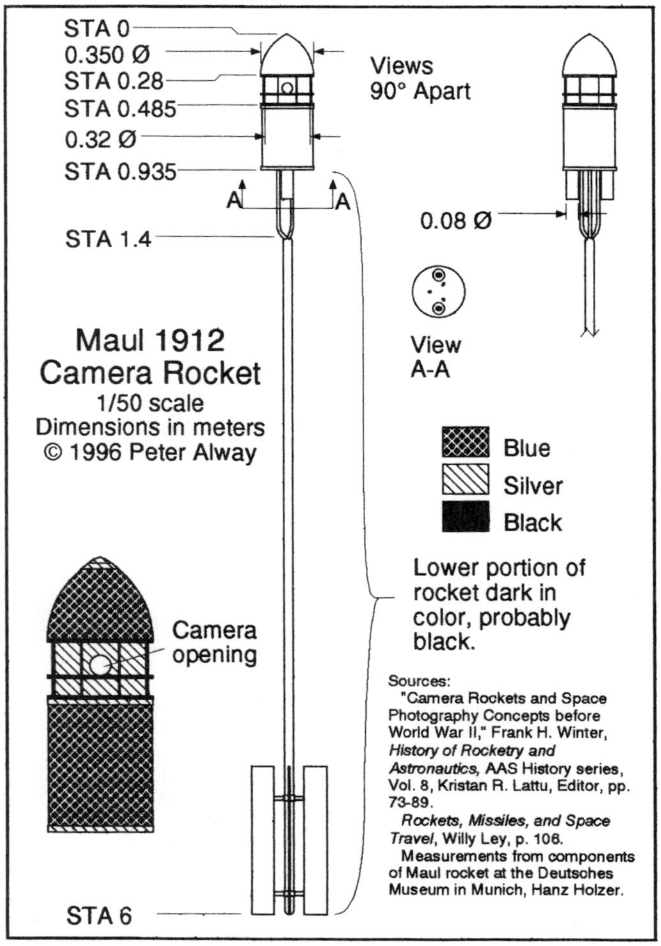

Alfred Maul's 1912 Rocket Specifications

Weight	42 kg (92 lb)
Altitude	800 m (2,600 ft)
Length	6 m (20 ft)
Diameter	320 mm (12.6 in)

Reinhold Tiling's Rockets

The second significant solid rocket experimenter was Reinhold Tiling, an engineer in Osnabrük, Germany. Supported by Count von Ledebour, Tiling created and tested his rockets in private. Then on April 15, 1931, he offered a public demonstration of two impressive rocket types. The first was a rather conventional-looking powder rocket, about 1.5 m (5 ft) long. One example of this type reached an altitude of 2 km (1 1/4 mi). The second type anticipated the Space Shuttle's glide recovery. It launched vertically, with four long fins trailing behind. But shrouded in two fins were long wings, which deployed jackknife-style at apogee. The mechanism was effective, at least for flights in still air. These rockets reached from 500 to 800 meters (1500-2500 ft). Tiling also patented a helicopter recovery device, in which four fins of a similar rocket all deployed at an angle, but he did not demonstrate this invention.

Tiling had no use for the spaceflight enthusiasts—at a meeting with the Verein für Raumschiffahrt, Tiling told members he was used to dealing only with nobility, and that a group that had trouble with simple vertical ascents did not interest him. He also mentioned that he would soon be sending a rocket across the English Channel.

On October 10, 1933, Tiling and two assistants were compressing gunpowder into 20-kg (40 lb) propellant grains, perhaps for his Channel-crossing rocket. As they worked into the night, the press exploded. Whatever injuries Tiling and his two assistants suffered might not have been fatal if Tiling had not stored a large quantity of powder in the same room. While all three survived the immediate conflagration, they died from the burns within a day.

Tiling High-Altitude Rocket Specifications

Propellant weight	12 kg (14.3 lb)
Payload weight	5 kg (11 lb)
Length	1.5 m (5 ft)

STA 0
0.350 Ø
STA 0.28
STA 0.485
0.32 Ø
STA 0.935
STA 1.4
STA 6
Views 90° Apart
0.08 Ø
View A-A
Camera opening
Blue
Silver
Black
Lower portion of rocket dark in color, probably black.

Maul 1912 Camera Rocket
1/50 scale
Dimensions in meters
© 1996 Peter Alway

Sources:
"Camera Rockets and Space Photography Concepts before World War II," Frank H. Winter, *History of Rocketry and Astronautics*, AAS History series, Vol. 8, Kristan R. Lattu, Editor, pp. 73-89.
Rockets, Missiles, and Space Travel, Willy Ley, p. 106.
Measurements from components of Maul rocket at the Deutsches Museum in Munich, Hanz Holzer.

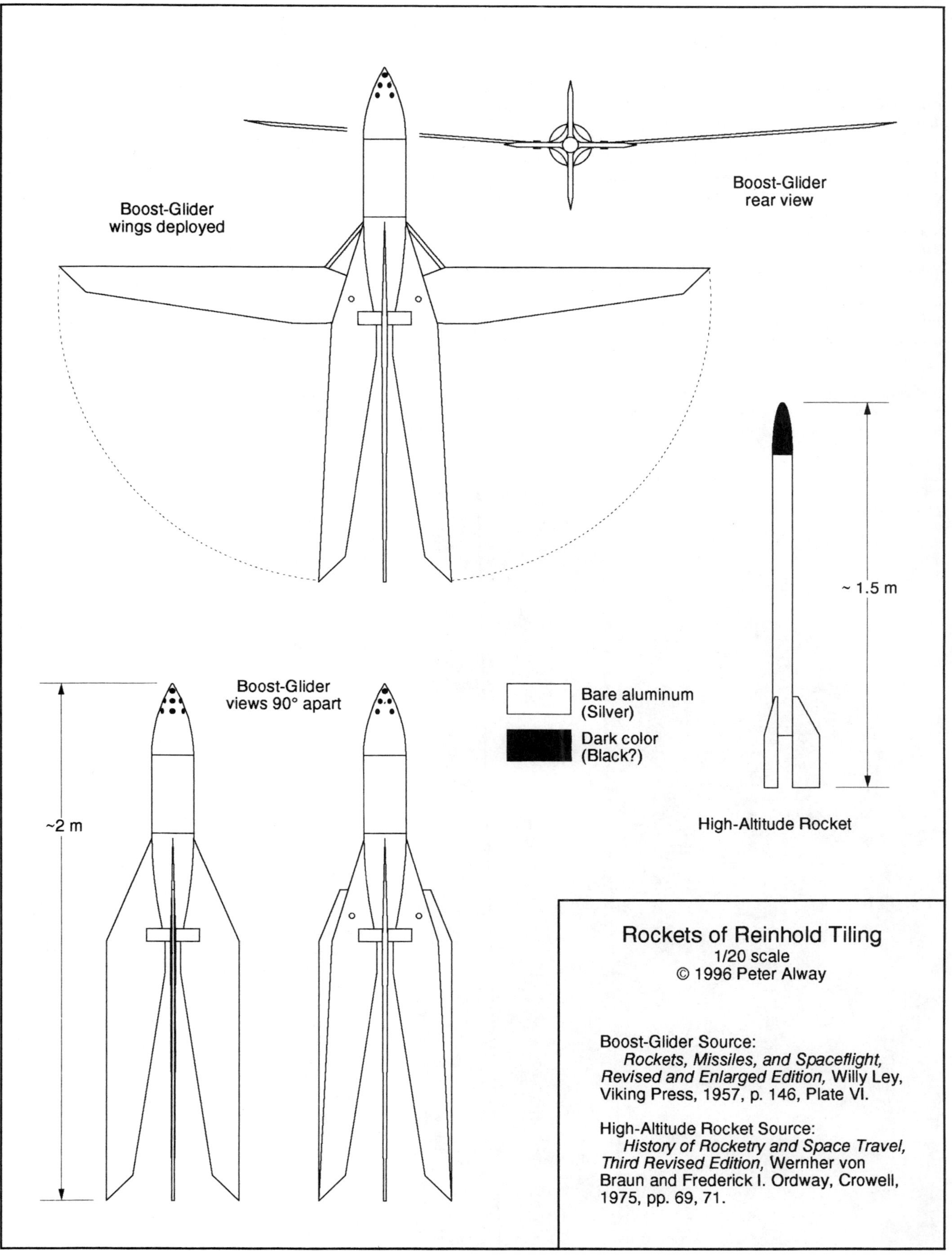

Boost-Glider
wings deployed
Boost-Glider
rear view
~ 1.5 m
~2 m
Boost-Glider
views 90° apart
Bare aluminum
(Silver)
Dark color
(Black?)
High-Altitude Rocket
Rockets of Reinhold Tiling
1/20 scale
© 1996 Peter Alway
Boost-Glider Source:
Rockets, Missiles, and Spaceflight, Revised and Enlarged Edition, Willy Ley, Viking Press, 1957, p. 146, Plate VI.
High-Altitude Rocket Source:
History of Rocketry and Space Travel, Third Revised Edition, Wernher von Braun and Frederick I. Ordway, Crowell, 1975, pp. 69, 71.

Johannes Winkler and the VfR

In 1923, Herman Oberth, a German-speaking Romanian mathematics professor, published *Die Rakete zu den Planetenräumen (The Rocket into Planetary Space),* in which he laid the foundations of astronautical theory. While Tsiolkovsky and Goddard had already worked out the theory of space travel independently, central Europe was unaware of their accomplishments. Oberth's dry, technical tract was published, and quickly became a popular success. Soon other Europeans—Frenchman Robert Esnault-Pelterie, Russian Nikolai A. Rynin, and Germans Max Valier and Walter Hohmann were calculating, writing and lecturing on space travel. Max Valier suggested the creation of a society for the advancement of space travel, and on July 5, 1927, Johannes Winkler, writer Willy Ley, and nine others founded the Verein für Raumschiffahrt—the Society for Space Travel (VfR).

Johannes Winkler was an engineer by training, but at the time of the VfR's founding he worked as a church administrator, which qualified him to handle the fledgling organization's bureaucratic problems. The group's first hurdle was linguistic; raumschiffahrt (literally, space ship travel) was not a German word, and German officials would not allow the group to incorporate as a registered society. Winkler eventually persuaded the appropriate officials that new inventions required new words, and that "raumschiffahrt" would be adequately defined in the society's charter.

The new organization attracted members from all over Germany and beyond. Valier, Hohmann, Oberth, Esnault-Pelterie, and Rynin joined.

In its first years, the VfR did not perform rocket experiments as an institution; instead, members performed research on their own. Valier performed some of the earliest tests, but much to the chagrin of other VfR members, he used commercial solid-fueled rockets to perform publicity stunts for the Opel auto company. Valier attached rockets to automobiles, train cars, and even a glider for Opel. In 1930, Valier turned to more serious work with liquid rockets, but he was killed when a combustion chamber exploded in a static test.

Oberth was the next member to engage in rocket experiments. In 1928 and 1929, Oberth worked for the UFA Film Company as a technical consultant for Fritz Lang's film, *Frau im Mond.* Oberth saw the film as a way to fund rocket experiments, and offered to build and launch a liquid-fueled rocket for the premier of the movie. A theoretically-inclined academic, Oberth advertised in the newspaper for an engineer. Rudolf Nebel answered, selling himself as an engineer. In time, Nebel would prove a far better salesman than engineer. Oberth's liquid-propelled rocket never flew—its streamlined metal casing and a wooden mockup became VfR mascots, decorating Society offices for years to come.

Johannes Winkler with his HW-I, Europe's first liquid-propelled rocket. (Smithsonian Institution photo No. A979A)

HW-1

In 1930, perhaps discouraged by the group's tendencies toward publicity-grabbing stunts, Winkler did much of his own rocket research independently of the VfR.

In 1929, the aircraft manufacturer Junkers hired Winkler to study solid propellant rockets for use in the jet-assisted takeoff of aircraft. In 1930, he left Junkers for a period to develop a liquid propellant rocket financed by the wealthy manufacturer and rocket enthusiast Hugo Hückel. Hückel simultaneously funded the VfR's operations as well.

Early in 1931, Winkler produced a simple liquid-propelled rocket, the HW-I (Hückel-Winkler I). A central combustion chamber was surrounded by three cylindrical tanks, one containing liquid oxygen, the second containing pressurized liquid methane, and the third pressurized nitrogen gas. The rocket, in the form of a triangular prism, bore little resemblance to a flying machine. The 5 kg

HW-I
Front view

Johannes Winkler's HW-I

1/8 scale
Dimensions in millimeters

Sources:
Rockets, Missiles, and Spaceflight, Revised and Enlarged Edition, Willy Ley, Viking Press, 1957, p. 145.
Smithsonian Institution photo no. A-979-A.
The History of Japanese Rockets: From Fire Arrows to Space Travel, Mita, 1996, ISBN 4-89583-150-7, p. 37.

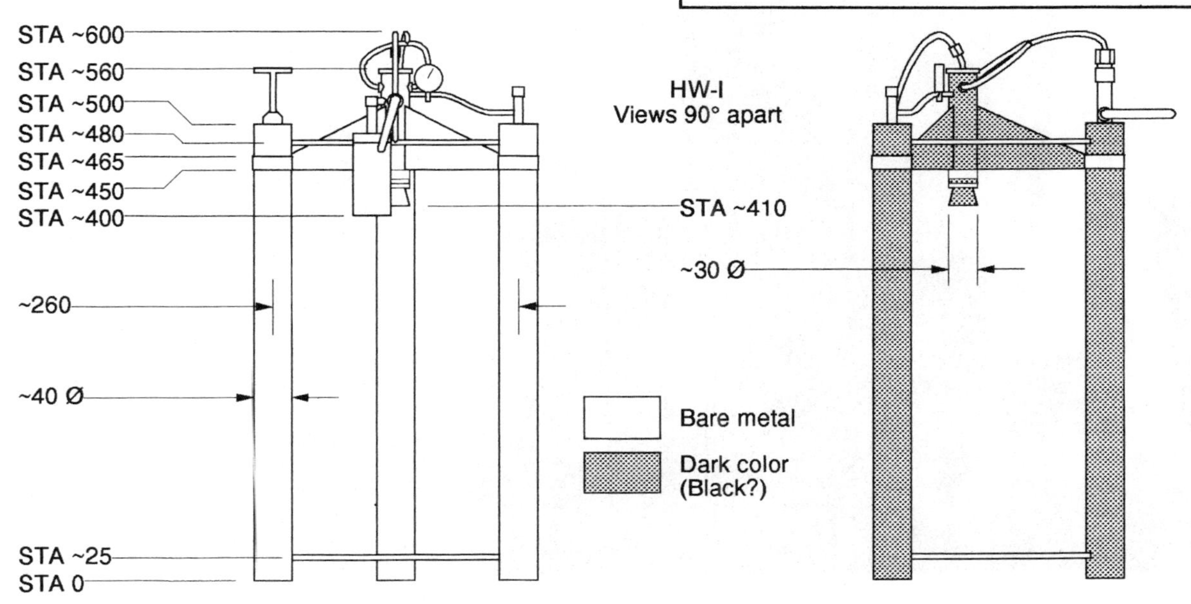

HW-I
Views 90° apart

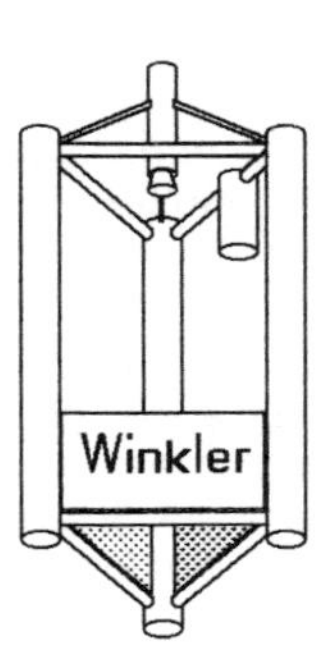

HW-IA
Sketch above shows general arrangement of major components. Engine is centered at top.

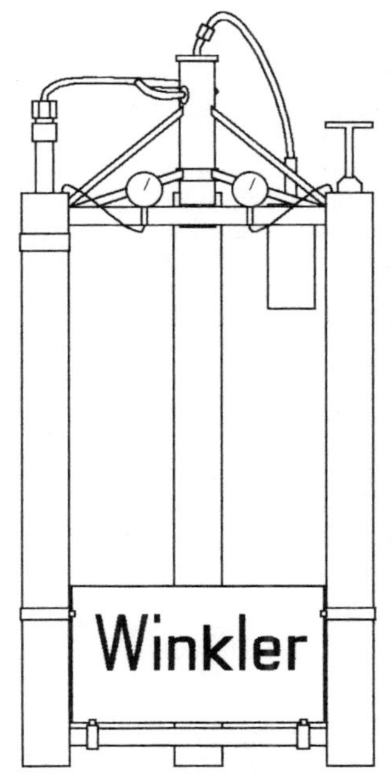

HW-IA
Note addition of fins between tanks, higher position of tanks, larger tubes between tanks, and different engine supports. This view is rotated 120° from the left view of the HW-1.

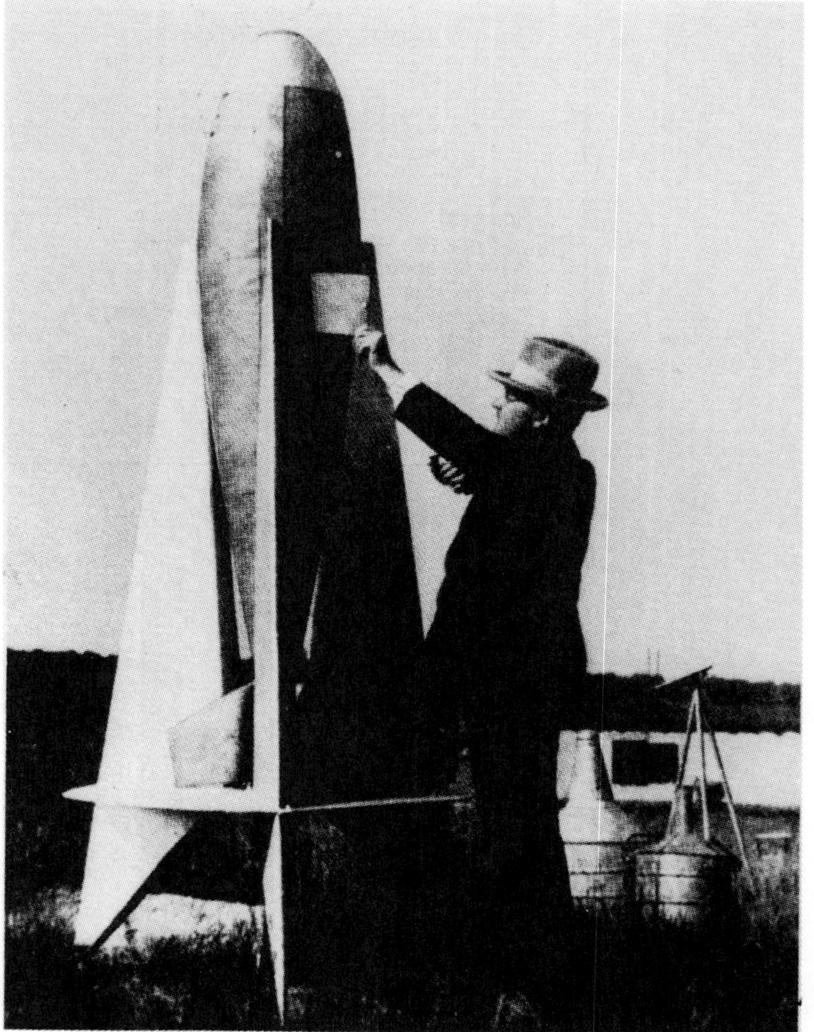

Winkler's HW-II was shrouded in a beautifully streamlined casing. The large delta "fins" and Jetsons-esque ring are actually parts of the launcher. (Smithsonian Institution Photo No. 76-16745)

(11-lb) machine carried about 1.7 kg (3 3/4 lb) of propellants. While the rocket hopped off the ground in a test on February 21, 1931, Winkler did not consider this a flight. The first "real" launch took place on March 14, 1931. Two weeks later, he described the flight:

> "By 4:45 pm, everything was ready for me to throw the ignition switch. It was an exalting and blissful moment when the apparatus rose from the launching table and climbed up with a rumbling, metallic hiss. Its motion was very steady. At a certain height, the apparatus turned over further and further into the horizontal, then maintained this direction for some time and finally landed at a distance of nearly 200 m [700 ft] from the launching point."

The "certain height" of this test is not recorded, but it was less than the intended 500 m (1700 ft). Most of the world thought this was the first liquid fueled rocket to fly until 1936 when Goddard revealed his 1926 flight. Winkler also tested an improved version, the HW-IA. This rocket can be identified by three stabilizing surfaces supported box kite-style between adjacent tanks.

HW-I Specifications

Launch weight	5 kg (11 lb)
Empty weight	3.3 kg (7 lb)
Expected altitude	500 m (1700 ft)
Length	1.5 m (60 in)

HW-II

In the summer of 1932, a second design was ready to fly. Winkler built the HW-II for maximum efficiency. Seventy-two percent of the rocket's mass was propellant, a record for the time (including Goddard's rockets), and a teardrop-shaped shell minimized aerodynamic drag. Like the HW-I, the new rocket burned methane and liquid oxygen. The latest technology was used throughout, including propellant valves made of "elektron," a new aluminum-magnesium alloy.

Winkler was not one for extensive testing. His faith in the laws of physics justified a naive view about temperamental rocket engines; "if it works once, it will always work—at any rate, for me."

After some resistance from safety-minded officials, Winkler obtained permission to launch the rocket from the Frische Nehrung on the sea coast of East Prussia. But here the new "elektron" alloy was exposed to salt sea air, and began to corrode. Winkler and his assistant, Rolf Engel did not discover the leaky propellant valves until the beach was cleared, the sea was cordoned off, and government observers were on the way. They hurriedly flushed the shell of the rocket with nitrogen, hoping to eliminate the risk of explosion. When the ignition switch was fired, the rocket leapt from the pad, but in a fraction of a second, the remaining gas inside the aerodynamic shell ignited, blowing the rocket to bits. Winkler gave up on independent rocket research and returned to work at Junkers.

In addition to his experimental work, Johannes Winkler theorized about a unique alternative to building gigantic space rockets. He proposed building rockets from small standard modules that could be stacked or clustered as needed to propel a rocket into space. Thirty years after Winkler's 1947 death, German experiments created the OTRAG rocket along these lines. They tested small versions of their rocket with a few standard units, but in the 1980's, technical and political problems ended the effort.

HW-II Specifications

Launch weight	46 kg (100 lb)
Empty weight	10 kg (22 lb)
Thrust	940 N (210 lb)
Duration (expected)	49 sec
Total impulse	46,000 N-s (10,400 lb-s)
NAR designation	P 940
Length	1.90 m (6 ft 2.8 in)
Diameter	40 cm (15.75 in)

Johannes Winkler's HW-II

1/10 scale
Dimensions in Millimeters
Drawn by Peter Alway

Sources:

"A Man of the First Hour—Johannes Winkler," by Rolf Engel, *History of Rocketry and Astronautics,* AAS History Series Volume 10, A. Ingemar Skoog, ed., American Astronautical Society, 1990, pp. 271-284.

Drawing by Tasillo Römisch from the Space Model Museum of Mittweida/Saxony.

The Cambridge Encyclopedia of Space, ed. by Michael Rycroft, inside front cover.

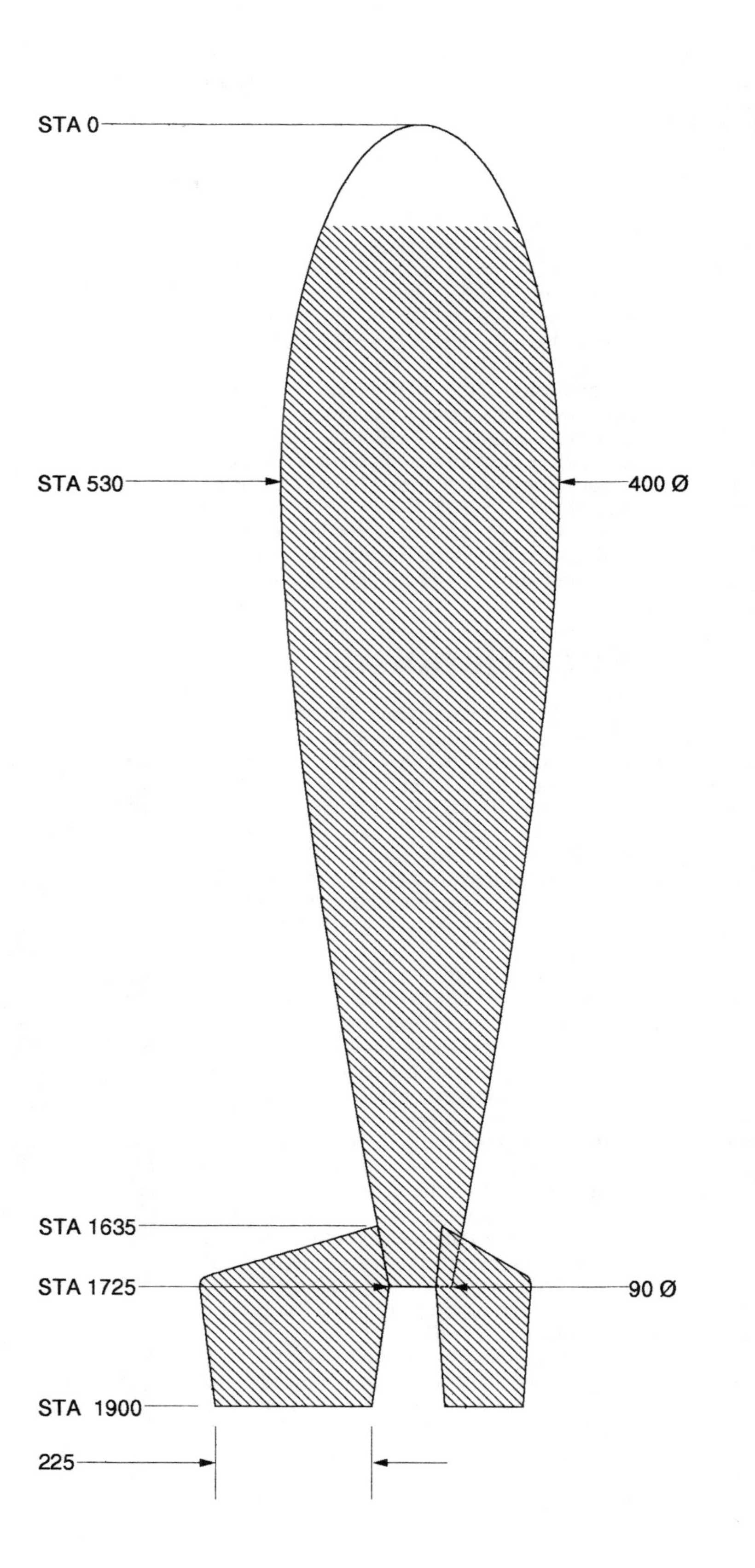

White

Silver

The VfR at the Raketenflugplatz

In 1930, the VfR changed location and changed direction. Winkler, the founding president, resigned to pursue his work with Junkers. Oberth, along with engineer Klaus Riedel and Wernher von Braun, an engineering student, set about turning Oberth's liquid rocket combustion chamber into a functioning engine. Nebel suggested that the group begin work with the simplest possible liquid-fueled rocket, a minimum rocket, or "Mirak." Riedel, Nebel, and others, including von Braun tinkered with the Mirak in 1930 and into 1931. The Mirak (three versions were built) could produce enough thrust to fly in static tests, but it never flew.

Early VfR rockets were cooled by immersing the combustion chamber in liquid. Oberth's combustion chamber was immersed in liquid oxygen. When this scheme seemed to trigger explosions, the VfR began to enclose combustion chambers in water jackets for cooling.

In August, Nebel located an abandoned army munitions dump on the outskirts of Berlin, and persuaded the city to lease it to the group for a token sum. Nebel christened the site "Raketenflugplatz" (rocket flying field), and the VfR set up business. Nebel set up a bachelor household in an abandoned building, and over time, an assortment of unemployed engineers, including Klaus Riedel, would do the same.

2-Stick Repulsor

In March of 1931, word came from Winkler that he had successfully launched a liquid-propelled rocket, believed at the time to be the first in the world. Riedel and Ley, disappointed that they had not been in on the flight, plotted their own experiment. Using components of an older version of the Mirak, they would assemble and fly their own rocket at the Raketenflugplatz. On May 10, Riedel took the rocket out for a static test. The rocket broke free, and crept up to an altitude of about 20 meters (60 ft). Riedel called Ley, and the two set about to repeat the experiment with a better model.

Repulsor No. 1, named for a 19th century science fiction spaceship, consisted of a rocket chamber immersed in a water jacket, and two tubular propellant tanks trailing behind on either side. Like Goddard years before, Riedel and Ley placed the engine at the front of the rocket in a tractor design. They thought that a forward engine would lead the rocket in a straight path like a child pulling a wagon. But Goddard had already recognized the fallacy of the tractor design: unlike the child pulling the wagon, the force of the rocket was applied in the direction the vehicle was pointed, and not in the direction of intended motion. Stick-stabilized rockets fly straight not because they are propelled from the front, but because the heavy pyrotechnics are ahead of the lightweight stick. Without benefit of stabilizing fins, the Repulsor whizzed around the sky, crashing to the ground after looping as high as 60 meters (200 ft).

Ley and Riedel assembled the second Repulsor with more thought. They moved the tanks closer together, and added aluminum hoops at the bottom to maintain spacing. Ley and Riedel cut four fins from an aluminum disk, and bolted them to the hoops. Vapor boiling from the liquid oxygen generated 20 atmospheres of pressure to feed the liquid into the engine. Nitrogen gas in the fuel tank forced gasoline into the engine. The whole assembly cost no more than $8.00 (US). Ley described the launch on May 23, 1931:

> "While Riedel waited, wristwatch in hand, for the oxygen to build up pressure, I climbed a low hill to have a vantage point from which I could not only see the Repulsor, but also look over most of the Raketenflugplatz. From down below I heard the shouts of 'Ready!—Fire!' and saw a white flame shoot out of the nozzle. It shortened quickly, became less bright, and began to roar. At the same instant, the Repulsor rose off the ground, slowly at first, but accelerating rapidly. It climbed to about 60 meters (200 ft) and then turned sideways, like a car taking a curve. In that position it raced across the whole place, still under full power. It slowly lost altitude and it also slowly turned around its longitudinal axis while in flight. The four aluminum fins reflected the red rays of the setting sun like mirrors."

VfR members chased the rocket off the Raketenflugplatz grounds to a tree 600 meters (2000 ft) from the launch site, where the wreckage hung 30 feet off the ground.

With that flight, Mirak was dead. The VfR was, after all, an amateur group, and actual rocket flights were far more gratifying than static tests. Riedel and Ley built and flew four more 2-stick Repulsors in June and July of 1931. These "Repulsor No. 3" rockets included parachutes, allowing recovery for repeated flights. One 2-stick Repulsor reached an altitude of about 600 meters (2000 ft).

2-Stick Repulsor No. 2 Specifications

Thrust	250 N (55 lb)
Range	600 m (2000 ft)
Length	~1.5 m (5 ft)

Klaus Riedel and Repulsor No. 2, a 2-Stick Repulsor with fins. (Smithsonian Institution negative no. A-4195-C)

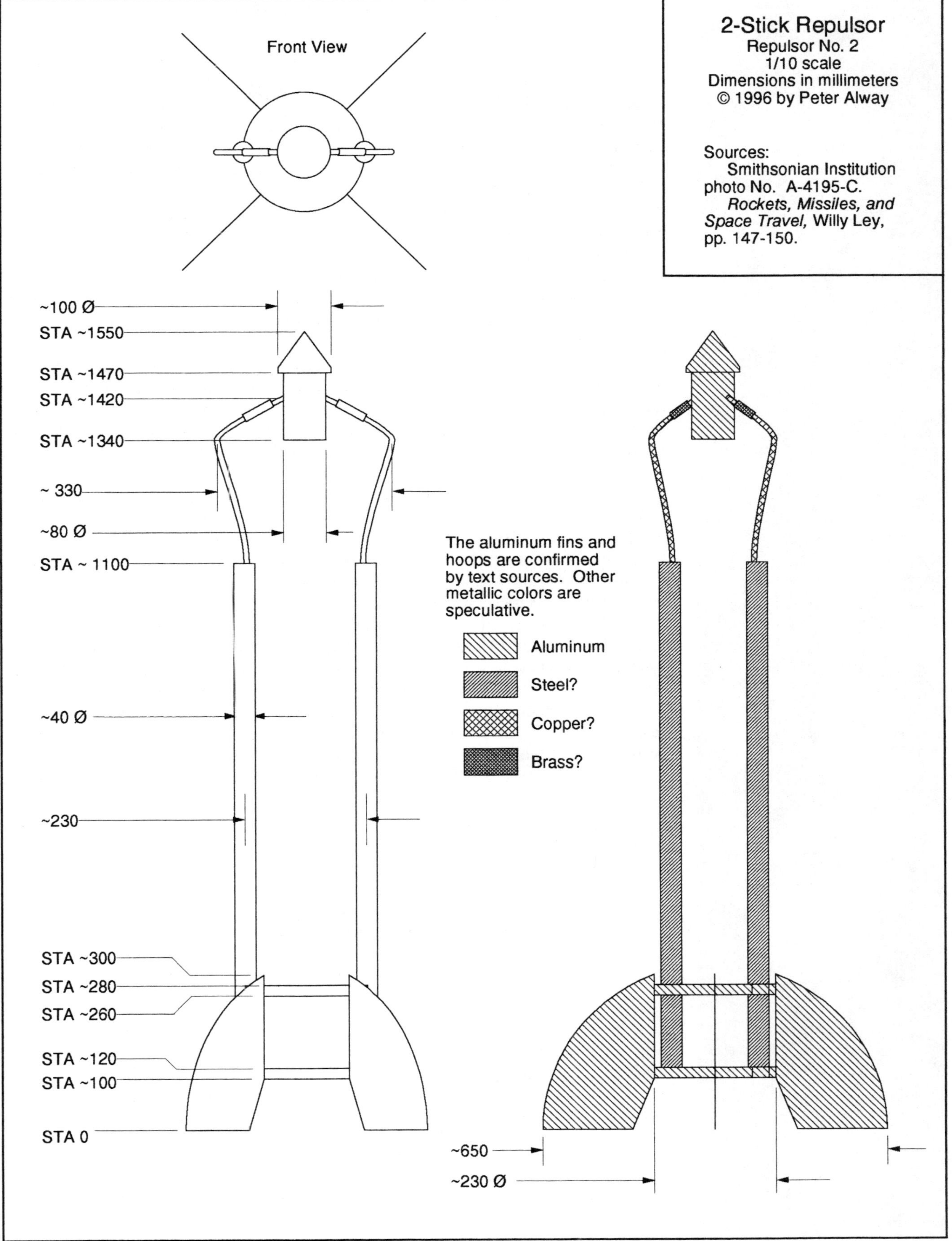
2-Stick Repulsor
Repulsor No. 2
1/10 scale
Dimensions in millimeters
© 1996 by Peter Alway
Sources:
Smithsonian Institution photo No. A-4195-C.
Rockets, Missiles, and Space Travel, Willy Ley, pp. 147-150.
Front View
~100 Ø
STA ~1550
STA ~1470
STA ~1420
STA ~1340
~ 330
~80 Ø
STA ~ 1100
~40 Ø
~230
STA ~300
STA ~280
STA ~260
STA ~120
STA ~100
STA 0
The aluminum fins and hoops are confirmed by text sources. Other metallic colors are speculative.
Aluminum
Steel?
Copper?
Brass?
~650
~230 Ø

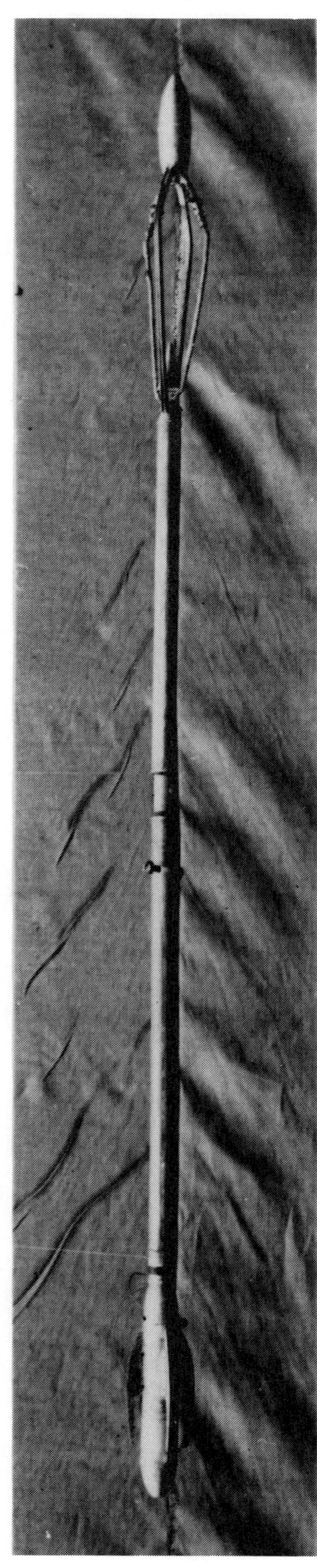

1-Stick Repulsor

In August of 1931, VfR members rearranged the Repulsor with the propellant tanks in line behind the combustion chamber, resulting in Repulsor 4, a "one-stick Repulsor." The engine was encased in a streamlined water jacket, supported in front of the oxygen tank. Behind the oxygen tank was the gasoline tank, and behind that was the parachute compartment. Small fins attached to the parachute compartment provided aerodynamic stability. The first flight reached an altitude of about a kilometer, and was successfully recovered by a parachute ejected from the tail The VfR built several of these machines, including two longer rockets with extended tanks. By May of 1932, VfR members boasted of 87 Repulsor flights to altitudes of up to about 1.5 km (1 mi) and ranges (by accident) as great as 5 km (3 mi).

One flight, in front of newsreel cameras, flew into a shed owned by the local police station, and burned it to the ground, nearly ending the VfR's experiments. While static tests continued at the Raketenflugplatz, future significant launches would take place at remote sites.

In the Spring of 1932, the VfR put on a show for officers of the German Army. The Army representatives were disappointed that experimenters could not produce data on engine performance, fuel consumption, heat transfer, or pressure. Nebel, flush with visions of a huge Army research contract was disappointed to receive only funds to build a one-stick Repulsor for an 8-km (5 mi) high demonstration flight at Kummersdorf in July.

The Army demonstration version of the 1-stick Repulsor carried both a parachute and a flare to be ejected at altitude. The silver-painted rocket was powered by gasoline and liquid oxygen. Pressurized nitrogen forced gasoline into the combustion chamber at the front of the rocket. The engine was cooled by a water jacket. Nebel scaled back his claims to an altitude of 3.5 km (about 2 miles)—still higher than any previous flight, but conceivable if this rocket had larger tanks than the earlier Repulsors.

Nebel, von Braun, and Riedel took the rocket to Kummersdorf. The rocket flew straight up for just 30 meters (100 ft) before arcing off into nearby woods without ejecting a parachute or a flare. The peak of the arc was just 60 meters (200 ft). The Army officers were not especially impressed by the VfR's unreliable toylike rockets and unscientific attitude. Because of Nebel's declining performance predictions, the Army suspected that the failure was no honest malfunction, but that Nebel had exaggerated the performance of the rocket. The Army was, however, impressed by one VfR member. On October 1, they hired Wernher von Braun to study the problem of the long-range missile.

The VfR would take on one more project before its collapse—the notorious Magdeburg Rocket.

1-stick Repulsor Specifications
(Kummersdorf Demonstration)

Launch weight	12 kg (26 lbs)
Empty weight	7.12 kg (15.7 lbs)
Thrust	250 N (55 lb)
Duration	21 sec
Total impulse	5250 N-sec (1200 lb-sec)
NAR Designation	M 250
Altitude	600 m (2000 ft)
Length	4 m (13 ft 7 in)
Diameter	6 cm (2.36 in)

The VfR's 1-Stick Repulsor. Components are, from front to back: Combustion chamber immersed in water jacket, support struts and propellant lines, Conical blast shield, liquid oxygen tank, gasoline tank, parachute canister with fins. (Smithsonian Institution negative no. 77-7655)

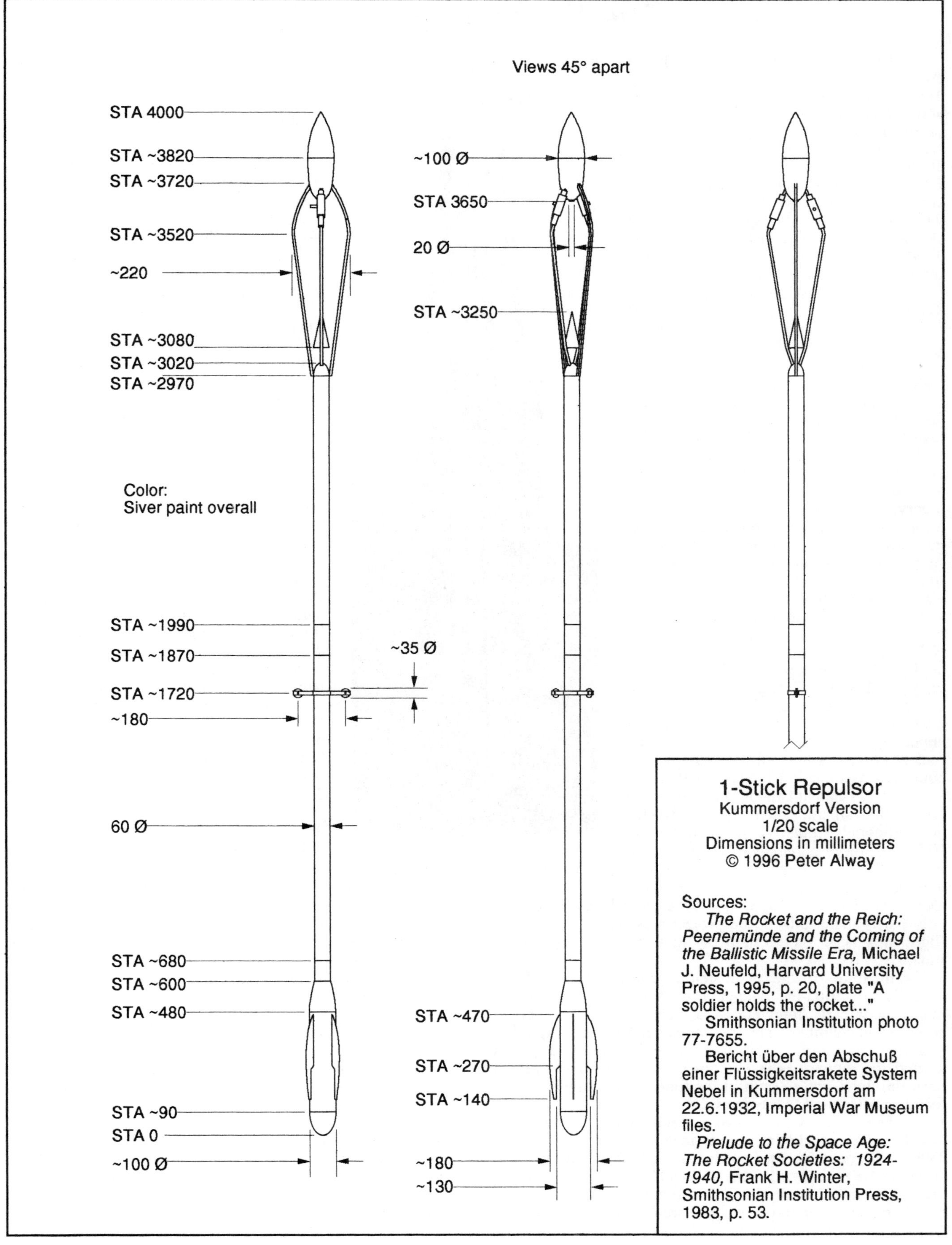

1-Stick Repulsor

Kummersdorf Version
1/20 scale
Dimensions in millimeters

Sources:

The Rocket and the Reich: Peenemünde and the Coming of the Ballistic Missile Era, Michael J. Neufeld, Harvard University Press, 1995, p. 20, plate "A soldier holds the rocket..."

Smithsonian Institution photo 77-7655.

Bericht über den Abschuß einer Flüssigkeitsrakete System Nebel in Kummersdorf am 22.6.1932, Imperial War Museum files.

Prelude to the Space Age: The Rocket Societies: 1924-1940, Frank H. Winter, Smithsonian Institution Press, 1983, p. 53.

The Magdeburg Project

Late one night in 1869, an aspiring American alchemist, Cyrus Ray Teed, had a vision; a beautiful woman came to him and told him two things. First that he was destined to be the new messiah, and second that the Earth was hollow, and we all live on the inside. Under the name Koresh, Teed founded a religion and concocted a concave cosmology which vaguely explained the appearance of the sky in terms of the inside of a sphere. Teed died in 1908, but his sect survived through the first half of the century (in the end, the last of his followers insisted that Teed's alchemy had anticipated the atomic bomb), and inspired a similar group in Germany, led by Peter Bender.

Bender, a badly wounded Luftwaffe pilot from World War I, expounded the Hollow Earth Doctrine, and Karl E. Neupert wrote *Geokosmos,* a textbook for followers. The idea appealed to Franz Mengering, an engineer in the city of Magdeburg. He proposed sending a rocket on a vertical ascent to test the theory; if the rocket landed at Magdeberg's antipode, the theory would be proven. Remarkably, Mengering successfully appealed to the civic pride of Magdeburg's government. If nothing else, a spectacular manned rocket ascent was appealing as a tourist attraction.

Mengering approached Nebel, and Nebel, smelling money, hastily agreed to apply "his" Raketenflugplatz to the project. The VfR's engineers, impressed by unemployment statistics so dismal that many Germans found even Adolf Hitler appealing, were happy to work on any rocket, no matter how silly its purpose. A contract was signed in January of 1933.

Nebel had taken on an impossible task; while all the talk of hollow worlds and antipodal flights was abandoned, the VfR, or more precisely, a group of Raketenflugplatz engineers who "happened" to be VfR members, could not possibly produce a manned rocket ascent by Easter of 1933. As time passed, and financing only partly materialized, the project was scaled back to the flight of a large model, and the launch was delayed until June.

In spite of the inauspicious circumstances of the rocket's creation, the Magdeburg rocket represented one real technical innovation—a regeneratively cooled combustion chamber. The fuel, an alcohol-water mix, rather than the accustomed gasoline, passed through a jacket around the engine, both to cool the chamber and prevent burn-throughs, and to preheat the fuel. Development of this feature contributed to the delays.

The rocket was clad in a metal casing for theatrical effect, but the rocket's performance was not so effective. Twice, the rocket refused to clear its launcher. On the final attempt, the rocket slid from its launcher, smashed into the ground, and skidded along 30 feet (10 m). The VfR members picked up the pieces and went home.

The 4-Stick Repulsor assembled from parts of the failed Magdeburg rocket. (Smithsonian Institution negative no. A-4195)

This was not the end of the rocket. VfR members scrapped the smashed decorative casing and rearranged the components into a functioning rocket. Thanks to the police shed fire at Raketenflugplatz, the rocketeers were forced to fly from Lindwerer (Lovers' Island) in Tegeler Lake, near Berlin.

The new incarnation of the Magdeburg rocket, called rocket 10-L, took the form of a four-stick Repulsor. Four cylindrical propellant tanks, alternating between liquid oxygen and an alcohol-water mix, were attached below the rocket's streamlined engine. Finlike vanes were attached, inexplicably, to the nose cone.

On July 14, just two weeks after the conclusion of the Magdeburg fiasco, the 10-L rocket was ready to fly. Remarkably, the rocket reached an altitude of 600 meters (2000 ft) before looping about the sky. The rocket splashed down 100 meters (300 ft) off shore, saved at the last instant by a late parachute deployment. After a second flight to 100 meters (300 ft) a week later, the island's owner kicked out the rocketeers—their contraptions were frightening the tourists.

In August, the workers from the Raketenflugplatz attempted three flights of this design from a police boat in Schwielow Lake. Each failed. In September, they adapted the rocket into a 2-stick Repulsor. Both launches failed.

As the 10-L rocket failed, so did the VfR. Unknown to its members, the VfR had already declared bankruptcy, thanks to a secret move by Nebel. By the end of 1933, the VfR broke up in a storm of allegations and counter allegations. On top of this, leaking pipes at the Raketen-flugplatz had run up a huge water bill. The rocket makers were sent packing, and the VfR vanished into history. The Army took over the site, and VfR's collection of tools and components (including Oberth's beloved rocket casing) was confiscated, never to be seen again.

4-Stick Repulsor Specifications

Launch weight	125 kg (275 lb)
Empty weight	70 kg (154 lb)
Thrust	2450 N (550 lb)
Duration	32.5 sec
Total impulse	78,400 N-s (17,600 lb-s)
NAR designation	P 2450
Altitude	600 m (2000 ft)
Length	3 m (10 ft)
Diameter	30 cm (11.8 in)

Rocket 10L
4-Stick Magdeburg Repulsor

1/15 scale
Dimensions in millimeters

Sources:

"The Development of Regeneratively Cooled Liquid Rocket Engines in Austria and Germany," Irene Sänger-Bredt and Rolf Engel, *First Steps Toward Space,* AAS History Series Vol. 6, 1985, pp. 226-227.

Smithsonian Institution photo no. A-4195.

Prelude to the Space Age: The Rocket Societies: 1924-1940, Frank H. Winter, Smithsonian Institution Press, 1983, p. 189.

Drawing of tank end cap from National Air and Space museum archives, SI # 78-18814.

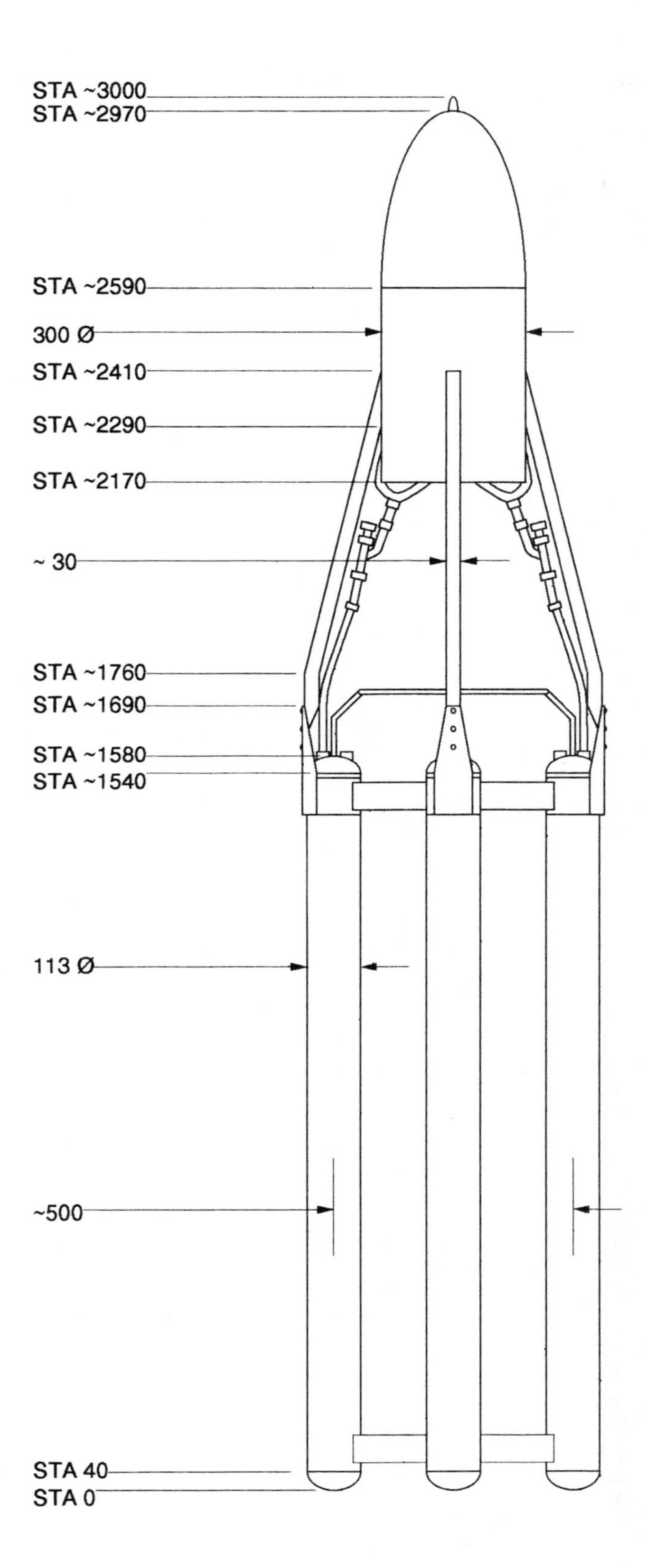

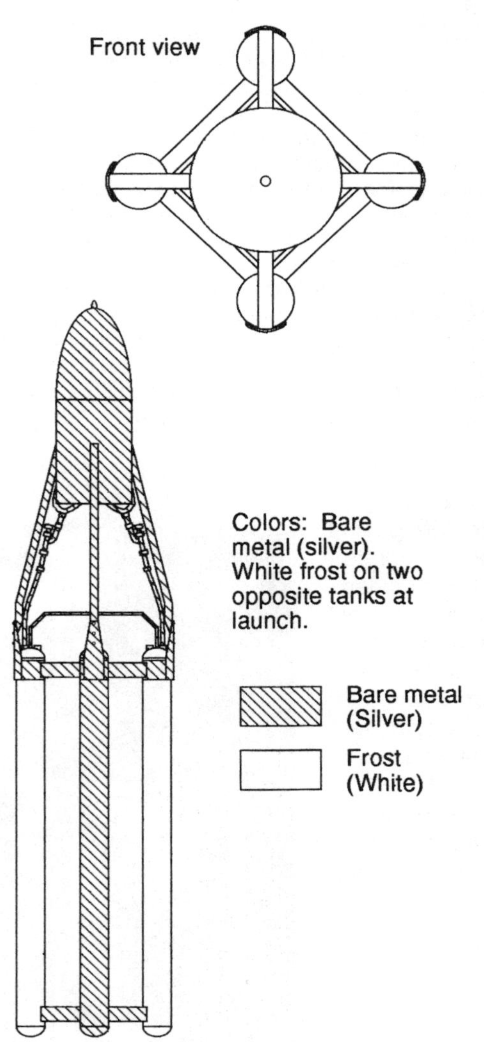

Wernher von Braun and the German Army

The A-2

In 1929, Karl Becker, an artillery expert with the Army Ordinance Office, began a program investigating solid-fueled rocket artillery. He soon expanded the program to include liquids. It was evident that there was little knowledge of liquid-fueled rocketry in German industry or academia. Becker may have supported the VfR's move to the "Raketenflugplatz," in hopes that the group might develop into a useful organization.

In 1931, Captain Walter Dornberger was charged with rounding up talent for the Army's liquid fueled rocket enterprise. Dornberger picked up staff wherever he could—from industry, academia, and even from the ranks of the amateur Verein für Raumschiffahrt (VfR). While the VfR's work left much to be desired (they perfomed static tests and flights as spectator events, without recording so much as a single thrust-time curve), some talented engineers stood out. Foremost among them was young Wernher von Braun, still an engineering student. After the 1932 demonstration of the 1-Stick Repulsor at Kummersdorf, Dornberger saw to it that von Braun would finish his graduate work with studies of rocket combustion and move to the Army upon graduation.

Late in 1932, von Braun began his rocket work for the Army at Kummersdorf with the help of Heinrich Grünow, a mechanic from the VfR. In 1933, von Braun created a 300-kg (660 lb) thrust engine burning alcohol with liquid oxygen. The engine was immersed in the alcohol tank for purposes of cooling. He incorporated the engine into a sausage-shaped rocket called the A-1 (for aggregate No. 1) which included a pressure feed for propellants and a large gyroscope for stability. This was a passive stabilizing system—the massive gyro in the nose, spinning at 9,000 rpm, would resist any change in the axis of the rocket. Von Braun described the test: "It took us exactly one half-year to build, and exactly one-half second to blow it up."

An A-3 during the extensive ground testing program at Kummersdorf. For flight, at least one rocket was covered with a coat of dark, glossy paint, reportedly dark grey. (Smithsonian Institution photo No. 77-14790)

In 1934, von Braun finished his PhD, and his slowly growing team built the A-2, in which the gyroscope was moved between the propellant tanks. The two A-2 rockets were nicknamed Max and Moritz after a pair of comic strip characters known in the US as the Katzenjammer Kids. On December 19, Max took flight, reaching an altitude of 1700 meters (5600 ft) before precession of the gyroscope set it tumbling. Moritz flew on December 20, also successfully. The success of the A-2 cleared the way for von Braun's missile development program.

A-2 Specifications

Launch weight	150 kg (330 lb)
Thrust	3000 N (660 lb)
Duration	16 sec
Total impulse	50,000 N-s (10,000 lb-sec)
NAR designation	P 3000
Length	1.4 m (4 ft 7 in)
Diameter	304 mm (12 in)

The A-3

With the success of the A-2, von Braun's group began work on an engine producing 1500 kg (3300 lb) of thrust. Again the engine was submerged in the fuel tank, and fed by alcohol and liquid oxygen under pressure. The guidance system, relying on gyroscopes and integrating accelerometers, was farmed out to a builder of marine navigational gyros. Vanes protruding into the rocket exhaust controlled the rocket. Long rotating rods connected the vanes to the guidance system in the nose. A tail ring reinforced the fins at launch, dropping away early in the flight.

It was during the design of the A-3 that Dornberger, von Braun, and Walter Riedel (no relation to the VfR's Klaus) laid out the specifications of the ultimate missile. The rocket would exceed the performance of the WW I's largest cannon, the "Paris Gun," which could fire 10 kg (22 lbs) to a range of 130 km (80 miles). The missile, the A-4, would send 1000 kg (2200 lbs) 260 km (160 miles).

On Dec. 4, 1937, after extensive ground testing, the first A-3 lifted off from a new rocket research center at Peenemünde, on the Baltic coast. As the rocket spun on its axis, the guidance system failed. Its gyroscopes' gimbals jammed, and a false signal was sent to release the parachute. A second launch on December 6 suffered the same fate. The remaining two rockets, launched on December 8 and 11, did not carry parachutes, but they weathercocked severely, indicating too much aerodynamic stability and too little control.

It was clear that the A-3 could not be the final step leading up to the operational missile. Because the name A-4 was already applied to the operational missile, the interim program was called the A-5.

A-3 Specifications

Launch weight	750 kg (1700 lb)
Thrust	15,000 N (3,300 lb)
Duration	45 sec
Total impulse	670,000 N-s (150,000 lb-s)
NAR designation	T 15000
Length	6.741 m (22 ft 1 in)
Diameter	673 mm (26.5 in)

A-2 and A-3

1/30 scale
Dimensions in millimeters

Sources:

A-2:
V-Missiles of the Third Reich, Dieter Hölsken. Monogram Aviation Publications, 1994, p. 17.

A-3:
The Rocket by David Baker, Crown Publishers, 1978, p. 36.
V2: Dawn of the Rocket Age, Joachim Engelmann, pp. 9, 11.
Color description provided by André Siewa (unidentified book excerpt).
"Agregat 3" German language drawing, dated 3 August 1942, Heeresanstalt Peenemünde (HAP) drawing No. 4919E.
V-Missiles of the Third Reich, Dieter Hölsken. Monogram Aviation Publications, 1994, pp. 15-19.
The Rocket and the Reich: Peenemünde and the Coming of the Ballistic Missile Era, Michael J. Neufeld, Harvard University Press, 1995, plates "An A-3, missing some of its exterior skin..." and "An A-3 is prepared for launch..."

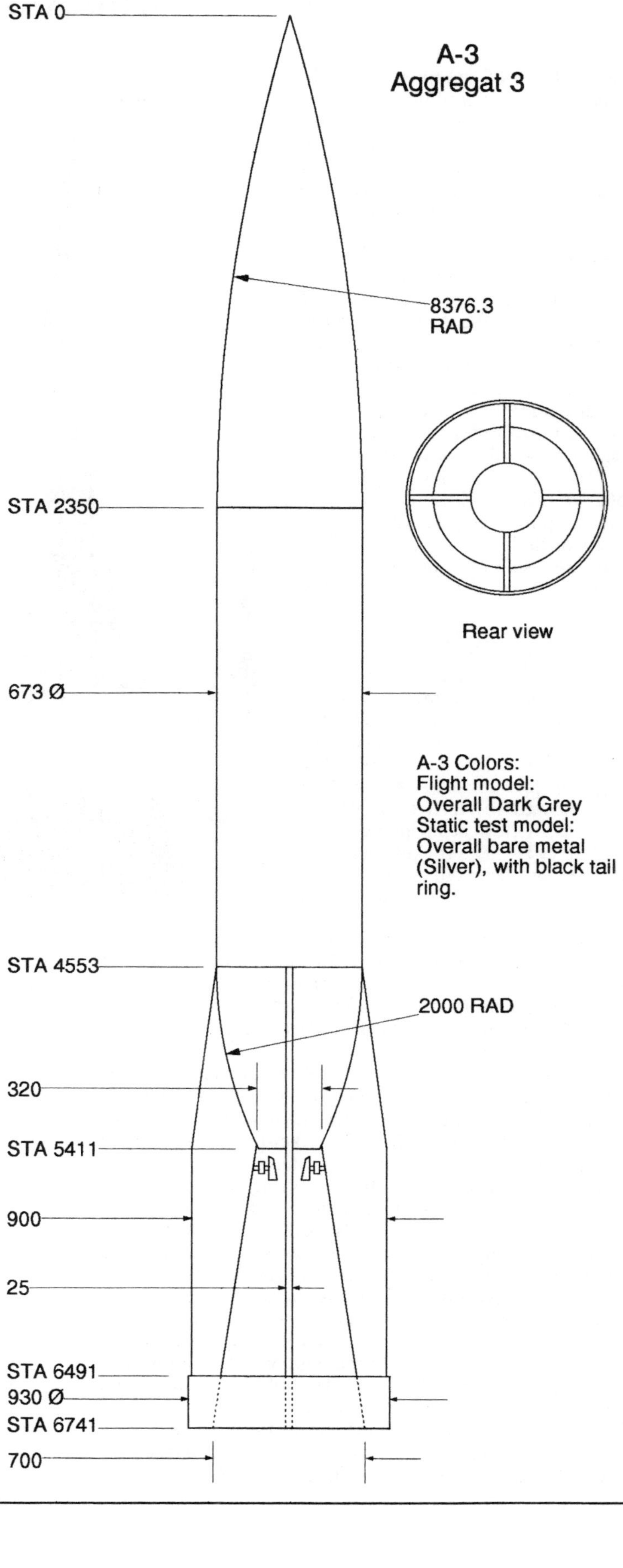

A-3
Aggregat 3

STA 0
8376.3 RAD
STA 2350
Rear view
673 Ø
A-3 Colors:
Flight model:
Overall Dark Grey
Static test model:
Overall bare metal (Silver), with black tail ring.
STA 4553
2000 RAD
320
STA 5411
900
25
STA 6491
930 Ø
STA 6741
700

A-2
Aggregat 2

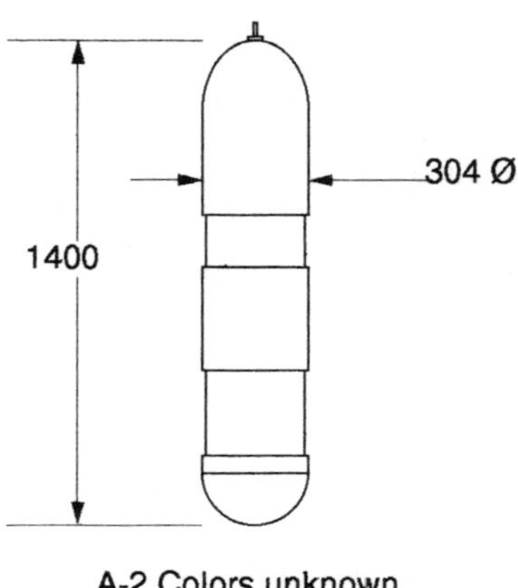

A-2 Colors unknown

The A-5

By 1938, von Braun's Peenemünde staff numbered in the hundreds. Their penultimate step on the way to the space rocket was the Aggregat No. 5, or A-5. The A-5 was an improved A-3 that would verify control and stability for the A-4 missile. It's propulsion system mirrored that of the A-3, with pressure-fed propellants and a combustion chamber enveloped by the alcohol tanks.

The fuselage of the A-5 resembled that of the A-3, but the fins were radically different. They were swept in accordance with recent discoveries in supersonic flight, and they were stubby so that when scaled up to a full-size missile, they could fit through a standard railway tunnel.

The first A-5's flew unguided in October of 1938. These four rockets, bearing a sporty red and yellow paint job, flew as high as 8 km (5 mi). They proved the stability of the design through most of the subsonic speed range.

The guidance system proved to be a challenge. The A-3 guidance system would be replaced by a system built by one of three competing contractors. The first of these systems was not ready until nearly 2 years after the A-3 flights. In October of 1939, one system flew aboard A-5's, and the others followed in April of 1940. At least 25 A-5's flew at Peenemünde by the end of 1941. The rockets deployed parachutes in their tail sections, and many were recovered from the Baltic to be flown again.

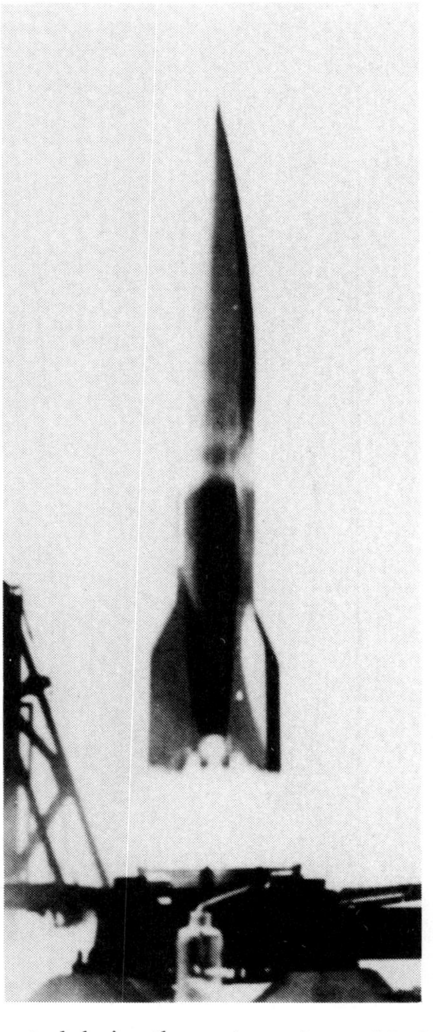

This later A-5 photo can be distinguished from a V-2 by the protruding jet vanes in the exhaust. (Smithsonian Institution photo no. 76-15523)

Because of the problems with roll control during the A-3 flights, the later, guided A-5's bore stark black and white roll patterns. This style of painting would continue through the A-4 program, and after World War II, into von Braun and company's American projects, up to and including the Saturn V Moon rocket.

Not all A-5 tests were full scale or powered rockets. A number of 20-cm (8") diameter, 1.6 m (63") tall models powered by Walter engines (by the manufacturer of the engines that powered the Me-163 rocket-powered aircraft) flew from Peenemünde beginning in March of 1939, testing fin configurations and possibly supporting an anti-aircraft missile program. The A-5 never reached supersonic speeds, and while supersonic wind tunnels were under construction, von Braun's aerodynamicists questioned the stability of the rocket near the speed of sound. To check this out, the Peenemünde shops produced iron replicas of the A-5 to be dropped from aircraft in the Spring of 1940. Observation of these unpowered drop-test models as they approached the speed of sound convinced airborne observers (including von Braun himself) that the final missile could break the sound barrier without flipping out of control.

Von Braun's Peenemünde group treated the A-5 as an all-around testbed. They tried out guidance for anti-aircraft missiles, and in one form, the A-7, it was even to fly with wings.

But the fundamental reason for the A-5 was development of A-4 aerodynamics, guidance, and control. On October 3 of 1942, on the third try, the A-4 lifted off, flew to an altitude of 80 km (50 mi) and a range of 190 km (120 mi). Because the rocket travelled some of the distance to its Baltic splashdown point above the sensible atmosphere, Dornberger justifiably declared the test to be space travel. When the A-4 came into use, the German Propaganda Ministry gave the weapon the V-2 designation. The V-2 was counted as the second of the Vergeltung (Vengeance) weapons, the V-1, sometimes called the "Buzz Bomb," being a pulse-jet cruise missile.

Notorious in war, both for the 2700 people killed by the missile and the 20,000 slave laborers who died in the Nordhausen factory where it was built, the V-2 would lead the way into space in the peace that followed. In 1945, the US shipped home parts for 100 German V-2 missiles, while almost all the top Peenemünde staff surrendered to the US Army and moved to America. The Soviet Union took over what was left of the facilities at Peenemünde and the underground V-2 factory at Nordhausen (after it had been thoroughly looted by the US Army). These men and machines were key to the beginnings of space travel in both superpowers.

In America, von Braun and over a hundred of his staff formed the core of Army Ballistic Missile Agency (ABMA) in Huntsville, Alabama. ABMA created the Redstone, Jupiter, and Pershing missiles. When NASA was born in 1958, the ABMA became the Marshall Spaceflight Center. Here the group developed the Saturn rockets, culminating in the Saturn V Moon rocket.

In the Soviet Union, the V-2 led to the R-1 missile, a nearly exact copy, and a series of rocket developments that led to the launch of Sputnik in 1957 and Yuri Gagarin's Vostok spacecraft in 1961.

A-5 Specifications

Launch weight	750 kg (1700 lb)
Thrust	15,000 N (3,300 lb)
Duration	45 sec
Total impulse	670,000 N-s (150,000 lb-s)
NAR designation	T 15000
Altitude	13 km (8 mi)
Length	5.825 m (19 ft 1 in)
Diameter	678.5 mm (26.7 in)

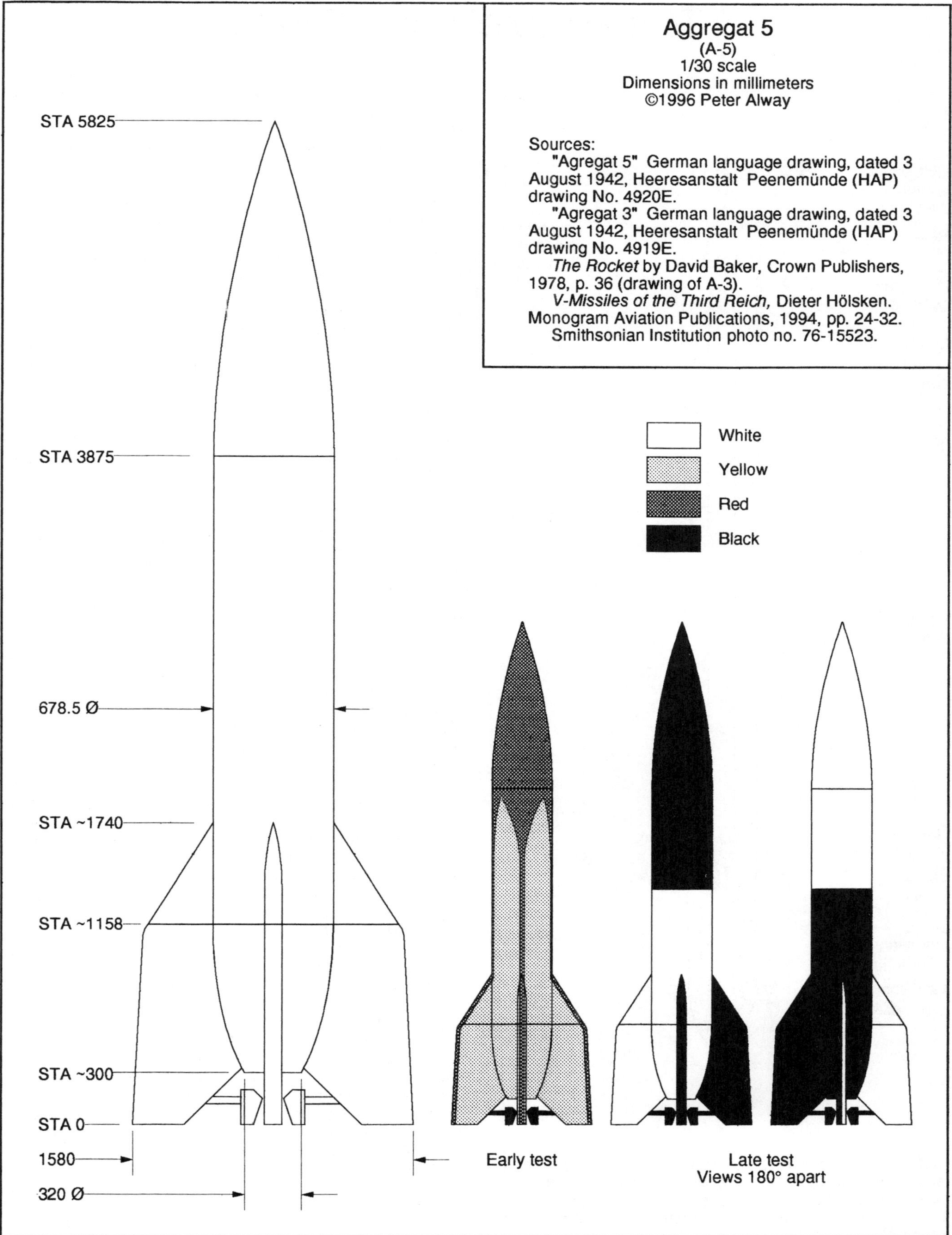
Aggregat 5
(A-5)
1/30 scale
Dimensions in millimeters
©1996 Peter Alway
Sources:
"Agregat 5" German language drawing, dated 3 August 1942, Heeresanstalt Peenemünde (HAP) drawing No. 4920E.
"Agregat 3" German language drawing, dated 3 August 1942, Heeresanstalt Peenemünde (HAP) drawing No. 4919E.
The Rocket by David Baker, Crown Publishers, 1978, p. 36 (drawing of A-3).
V-Missiles of the Third Reich, Dieter Hölsken. Monogram Aviation Publications, 1994, pp. 24-32.
Smithsonian Institution photo no. 76-15523.
STA 5825
STA 3875
678.5 Ø
STA ~1740
STA ~1158
STA ~300
STA 0
1580
320 Ø
White
Yellow
Red
Black
Early test
Late test
Views 180° apart

The Rockets of GIRD

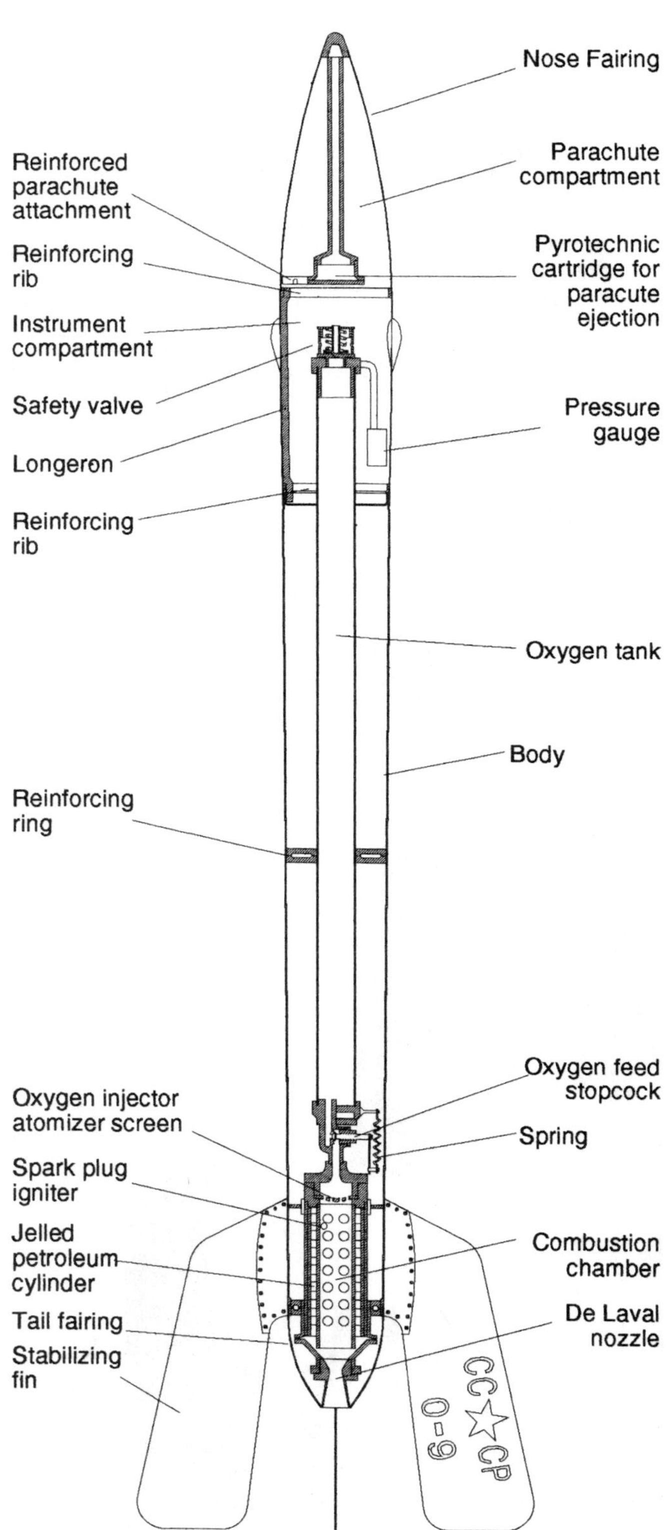

Components of the GIRD-09 rocket, the first of the GIRD rockets to take flight. The use of a solidified fuel with liquid oxygen made this the first hybrid rocket to fly. (diagram by author) Photo opposite shows GIRD-09 in museum exhibit. (NASA History Office)

In the Soviet Union, the transformation of astronautics from speculative theory to functioning metal fell to GIRD, variously translated as the Group for the Study of Reactive Motion or Jet Propulsion Study Group (Group Investigating Reaction Drive is not a bad approximation for those who would like to keep the acronym in English).

Konstantin Eduardovitch Tsiolkovsky was the elder statesman of astronautics in the Soviet Union. His early works anticipated those of Oberth and Goddard, but he never built a rocket. He did, however, inspire a generation of Soviet rocketeers. One of these was the theorist and charismatic space advocate Fridrikh Arturovich Tsander.

Like Tsiolkovsky, Tsander envisioned rocket-powered interplanetary craft based on rigorous mathematical analysis. Tsander's vision involved a mix of conventional aviation, jet propulsion, and rocketry. Tsander designed a biplane whose metallic wings, piston engines, and propellers would be melted in flight to become fuel for a liquid-propellant rocket engine.

Unlike Tsiolkovsky, Tsander worked on building rocket engines. In 1930, he converted a welder's torch into a small liquid rocket engine he called OR-1. In efforts to fund his schemes, Tsander approached the Society for Assisting Defense and Aviation and Chemical Construction in the USSR (Osoaviakhim). This government organization harnessed the energy and intellect of technical enthusiasts in the fields of aviation and chemistry. Distinctions between amateur and professional were not as clear in the Soviet Union as they are in the West; Osoaviakhim provided funding, even nominal salaries, yet it was considered a volunteer organization.

In 1931, Tsander found a friend at Osoaviakhim in the person of Ukrainian-born Sergei Pavlovich Korolev, an amateur glider pilot and professional aeronautical engineer with the Tupolev design group. Korolev saw that Tsander's spaceship was too far-out to attract government funding. Korolev and Tsander would have to develop a project that might be completed in months, rather than decades. They settled on a combination of Tsander's engine with a glider.

From this idea, Tsander and Korolev founded Moscow Group for the Study of Reactive Motion (Mos-GIRD—other GIRD branches formed later in other cities, but Mos-GIRD was the leader). For a workshop, Korolev quickly located an abandoned wine cellar he had used years before with a glider club. Endowed with Tsander's persuasive speaking voice, Korolev's engineering experience, a realistic project, and a place to build it, GIRD secured Osoaviakhim sponsorship.

In its first year, GIRD, led by Tsander, engaged in at least as much popularization ("technical propaganda") and theoretical talk as actual rocket construction. This was necessary to attract talent, and to educate these new engineers in rocket science.

In 1932, this would change. Sergei Pavlovich Korolev took over as chief of the organization. Korolev would turn it into a productive engineering organization. Korolev encouraged formal professionalism, and even extracted funding from Osoaviakhim to pay his engineers, machinists, and office workers minimal wages (before this, some members had noted the Russian acronym GIRD also stood for "Group of Engineers Working for Nothing").

Korolev made extraordinary efforts to scrounge the equipment needed for his rocket lab, soliciting a grinder from old glider friends at the Moscow Aviation Institute.

On another occasion, Osoaviakhim refused to supply GIRD with a desperately-needed lathe. Korolev outfitted his production chief with an Osoaviakhim uniform doctored to resemble that of a Red Army commander. Within days, he walked out of the Heavy Industry Commisariat, lathe requisition in hand.

Tsander was an inspiration to the members of GIRD; he devoted his full energies to the cause, working calculations late into the night, and often into the morning. Eventually, GIRD members established a strict policy of sending their patriarch home before closing up shop each night. This was certainly out of concern for Fridrikh Arturovich's health, but there may have been other motives. Infatuated with the idea of the consumable spaceship, Tsander distracted the engineers with problems of feeding powdered metal into the liquid propellant engines. Tsander would not be convinced that grinding wheels aboard a rocket would add more weight than they would save by powdering structure into fuel. Overworked and sickly, Tsander left Moscow for a "rest cure" early in 1933. He would not return.

GIRD-09

March of 1933 was a dark month for GIRD. Tsander's OR-2 engine blew up in its stand, and then Tsander himself died of typhus. The glider project was shelved. Deprived of their inspiring founder and their initial project, the rocket enthusiasts pursued a new direction. GIRD had examined several liquid propellant rocket projects, numbered 01 through 10. These were distributed amongst four engineering teams. GIRD-03 was to be a pump-fed engine for a glider. GIRD-05 was a nitric acid-kerosene rocket that would eventually come to fruition as the alcohol-oxygen "Aviavnito." GIRD 06 would be a winged rocket. GIRD-07 was a kerosene-oxygen rocket. This vehicle employed an engine-forward design with four cylindrical tanks trailing behind, concealed in four large fins. GIRD-08 was a ramjet project. GIRD-10 (GIRD-X) was Tsander's vertical ascent rocket project. On careful consideration, Korolev and his associates realized most of these were too complex for the novice rocket scientists. Project 09 showed the most promise for a first project. The others would continue, but at a slower pace.

Before turning to rocketry, Mikhail K. Tikhonravov had written numerous papers and a book on the flight of birds and ornithopters—flying machines with flapping wings. Tsiolkovsky's books had inspired Tikhonravov to design rockets even before GIRD's founding. Project 09 would be Tikhonravov's principal responsibility. At the heart of the GIRD 09 was an engine powered by jelled gasoline burning in liquid oxygen. An additive gave the semisolid form of gasoline the consistency of grease. It was chosen for its ability to form a thick layer insulating the combustion chamber walls from the heat of the flame. This propellant made a complex and heavy cooling system unnecessary. A special cylindrical grid held the charge of fuel in place in the combustion chamber.

Tikhonravov simplified the design by forcing liquid oxygen into the combustion chamber by its own vapor pressure. Because the fuel was acting as a solid propellant and the oxidizer was a liquid, the GIRD-09 would now be called a hybrid rocket, rather than a liquid-fueled rocket.

GIRD began work on Rocket 09 in earnest in March of 1933. Over the course of two and a half months, GIRD conducted 23 static firings, battling ignition failures, combustion chamber erosion, nozzle burn-throughs, ejection of unburned fuel and at least two explosions. By August of 1933, Tikhonravov and colleagues had adjusted their design enough to produce an engine that could produce 52.4 kg (116 lb) of thrust for 30 seconds.

GIRD researchers put considerable thought into the aerodynamic design of their first rocket, conducting wind tunnel tests on the rocket with various fin configurations at the Moscow Aviation Institute. The result was a sleek design with large stabilizing fins. Unlike Goddard's and Winkler's first liquid fueled "plumber's nightmare" rockets, the 8-foot tall GIRD-09 looked like the predecessor of the spaceship. It even came equipped with a payload compartment and a parachute recovery system.

On August 11, the rocket was loaded in a truck and taken out to the Nakhabino firing range outside of Moscow. A crowd of thirty GIRD members, a fraction of whom were needed for the launch, offered more advice than assistance as the GIRD production team loaded the 09 rocket onto its four-rail launcher.

The jelled petroleum filled the cells of a hollow cylinder of a metal honeycomb. Zina Kruglova rolled up her sleeves and loaded this cylinder into the combustion chamber. The rear closure of the combustion chamber held the fuel in place. N. I. Efremov, who had come up with the

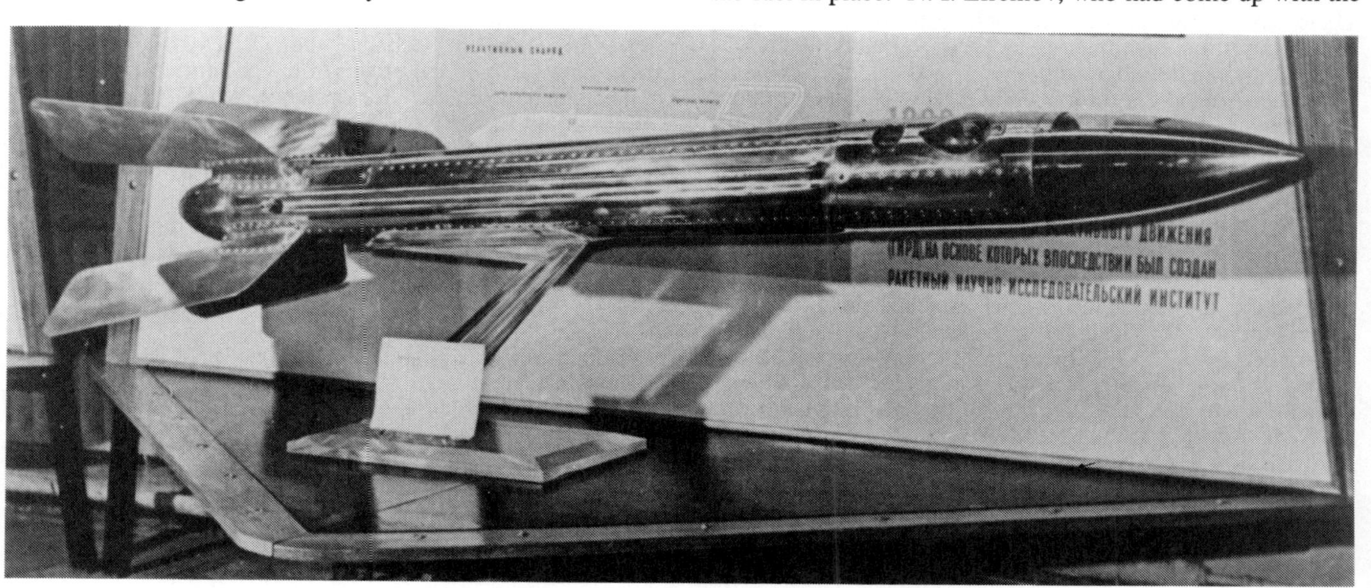

GIRD-09 awaits its flight in the woods outside Moscow. (Smithsonian Institution photo no. 82-389)

idea of jelled petroleum fuel, took the cleaner job of loading the liquid oxygen. After a 40-minute delay to fix a leaky valve, Korolev stood on tiptoes, matches in hand, watching the oxygen pressure gauge. When the oxygen pressure was adequate for launch, Korolev lit the delay fuse for the parachute ejection charge and ran for cover. One of the crew cranked the ignition generator. Nothing happened. There was a short in the wire running to the spark plug in the combustion chamber. A few seconds later, the parachute ejected with a bang.

On August 13, the crowd at a rainy Nakhabino range was smaller. This time flames erupted from the side of the combustion chamber, igniting the rocket casing. GIRD-09 went back to the shop for major repairs. On the way home, the personnel truck tipped, dumping the rocket team in the mud. GIRD's rocketeers began to fear they might really be crackpots.

On August 17, 1933, only 10 GIRD members traveled out to Nakhabino for the expected third failure of rocket number 09. The crew in the back of the truck carried the rocket safely across their laps as they rumbled along bumpy Russian country roads. Again, the crew slid the rocket onto its rails and filled it with fuel and liquid oxygen. As S. P. Korolev watched the oxygen pressure gauge, he saw it slow and stop before reaching full pressure. Somewhere a bit of ice had stuck a valve open, and the pressure would not increase. With low pressure, the GIRD-09 would fly, but not at optimum performance.

After a moment's consideration, Korolev lit the ejection fuse and ran for cover. The launch crew silently waited one carefully timed minute. Then Kruglova turned the generator connected to the spark plug. This time the plug sparked. At 7:00 pm, GIRD-09 slid off its 4-meter rails. Eyewitnesses all remarked on the slow liftoff and gradual acceleration of the rocket. The intended 2g acceleration of the rocket was much slower than that of more familiar solid-fueled rockets, and the low oxygen pressure may have slowed the liftoff further. The engine's 15-second burn was to carry the rocket 5 km (3 miles) into the air. But at 400 meters (1,300 ft), a flange at the base of the combustion chamber gave way, sending a jet of flame out the side of the rocket. The GIRD-09 turned on its side and shot into the woods. As the rocket plowed through the trees, it sheared off a fin, dented its casing, and broke in two. The flight was over in 18 seconds, well before parachute ejection.

In spite of its hard landing, the successful launch of a liquid-propelled rocket was a cause for celebration. The men and women of GIRD had proven themselves innovative engineers, not star-struck dreamers. Armaments Minister Mikhail N. Tukhachevski had long been pondering the transfer of GIRD from the volunteer Osoaviakhim to a new rocket research institute. On September 21, in light of the flight of GIRD-09, GIRD and the Gas Dynamics Laboratory (GDL) merged to form the Jet Propulsion Research Institute, RNII.

The GDL was an established professional military laboratory, secretly advancing solid fuel rocketry since 1921. The GDL had been examining liquid rocket engines since 1929. The GDL's Valentin Petrovich Glushko is acknowledged as the Soviet Union's great propulsion genius, but the GDL was strictly a laboratory organization. None of Glushko's engines had flown. Combined into RNII, the GDL's engineers could see their engines fly, and GIRD's engineers could enjoy professional pay and work .

As a part of RNII, the GIRD engineers built more GIRD-09 rockets, under the "Series 13" designation. The first of these flights on November 6 ended in an engine explosion just 100 meters (330 feet) off the ground. A January 1934 flight reached an altitude of 1500 meters (5000 feet). RNII flew six rockets in Series 13 at various elevation angles. While Korolev may have been personally interested in rockets as a means of space travel, military funding of rocket research dictated other directions. These last GIRD-09 flights were probably tests of the potential use of liquid propellants in artillery rockets.

In eighteen seconds, the GIRD-09 took Soviet rocketeers across the line from eccentric theorists to respectable engineers. Its designer, M. K. Tikhonravov, would go on to carry out the preliminary design studies for the Vostok program, which would launch the first human beings into space. Perhaps most importantly, the rockets of GIRD established the reputation of S. P. Korolev, who 24 years later would lead the group that placed the first satellite into orbit.

GIRD-09 Specifications

Launch weight:	18.95 kg (42 lb)
Payload weight:	6.2 kg (14 lb)
Propellant weight:	7.82 kg (17 lb)
Thrust:	514 N (115 lb)
Thrust duration:	15 sec
Total impulse:	7700 N-sec (1700 lb-sec)
NAR designation	M514
Length	2.457 m (8 ft 1 in)
Maximum diameter	178 mm (7 in)

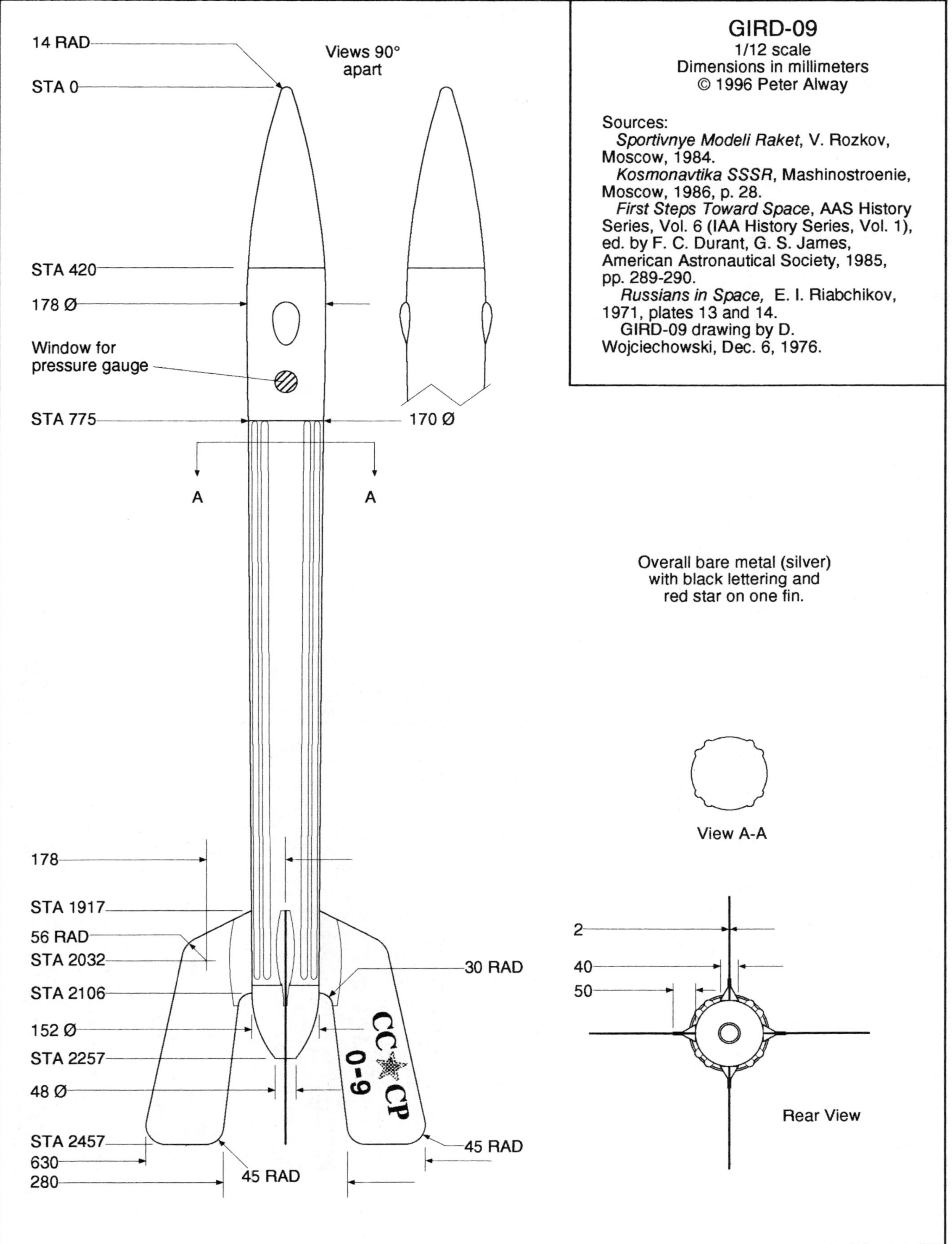
GIRD-09
1/12 scale
Dimensions in millimeters
© 1996 Peter Alway
Sources:
Sportivnye Modeli Raket, V. Rozkov, Moscow, 1984.
Kosmonavtika SSSR, Mashinostroenie, Moscow, 1986, p. 28.
First Steps Toward Space, AAS History Series, Vol. 6 (IAA History Series, Vol. 1), ed. by F. C. Durant, G. S. James, American Astronautical Society, 1985, pp. 289-290.
Russians in Space, E. I. Riabchikov, 1971, plates 13 and 14.
GIRD-09 drawing by D. Wojciechowski, Dec. 6, 1976.
Views 90° apart
14 RAD
STA 0
STA 420
178 Ø
Window for pressure gauge
STA 775
170 Ø
A
A
Overall bare metal (silver) with black lettering and red star on one fin.
View A-A
178
STA 1917
56 RAD
STA 2032
STA 2106
30 RAD
152 Ø
STA 2257
48 Ø
STA 2457
630
280
45 RAD
45 RAD
CCCP
0-9
2
40
50
Rear View

GIRD-X

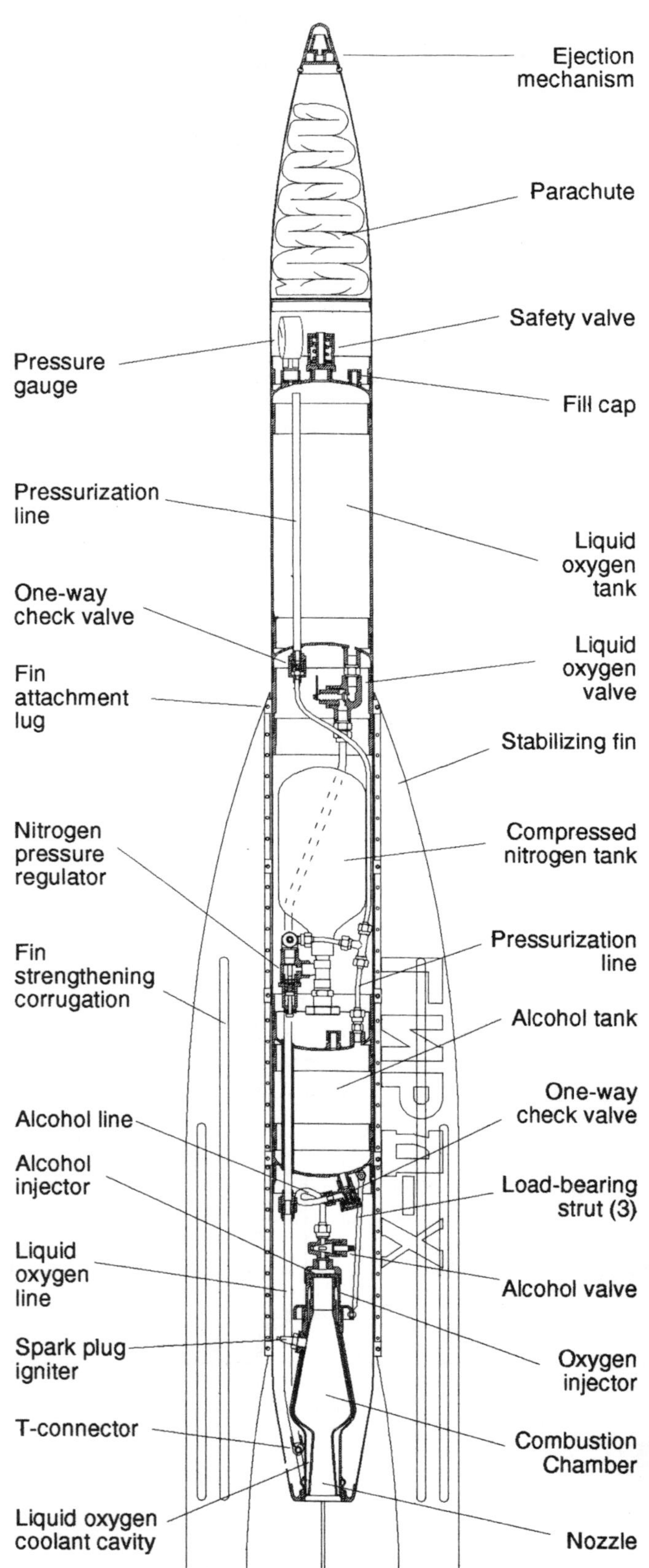

Components of the GIRD-X, the first entirely liquid-fueled Soviet rocket to fly. (diagram by author)

While GIRD's project 09 had enjoyed top priority since Tsander's death in March of 1933, Tsander's engineering group continued working on the OR-2 engine and project 10. In October, the OR-2 passed a static test, but it was the GIRD-X (Roman numeral 10) that would come to fruition in 1933.

GIRD-X would be propelled by a true liquid-propellant rocket engine. Pressure from a compressed nitrogen bottle would force both liquid oxygen and alcohol into the combustion chamber. To eliminate the weight of a thick refractory metal chamber, the GIRD-X engine would be regeneratively cooled—liquid oxygen passed through the hollow walls of the chamber and nozzle before entering the combustion chamber to react with the alcohol.

Just over 7 feet tall, the complete GIRD-X rocket was smaller than the GIRD-09, but it boasted slightly higher performance. On Nov. 25, 1933, the GIRD-X lifted off. It reached an altitude of just 80 meters before its combustion chamber burned through. Just as its predecessor did, the GIRD-X turned sharply but continued operating. It crashed under power 150 meters from the launcher. So ended the reign of the GIRD rockets. Further development would be under the auspices of the Jet Propulsion Research Institute (RNII).

GIRD-X Specifications

Launch weight:	29.5 kg (65 lb)
Thrust:	700 N (160 lb)
Thrust duration:	12 sec
Total impulse:	8400 N-sec (1900 lb-sec)
Length	2.2 m (7 ft 3 in)
Diameter	140 mm (5.5 in)

Members of GIRD with the GIRD-X rocket. Sergei Korolev is at the top left. (Novosti photo, also available as Smithsonian Institution no. 73-7133)

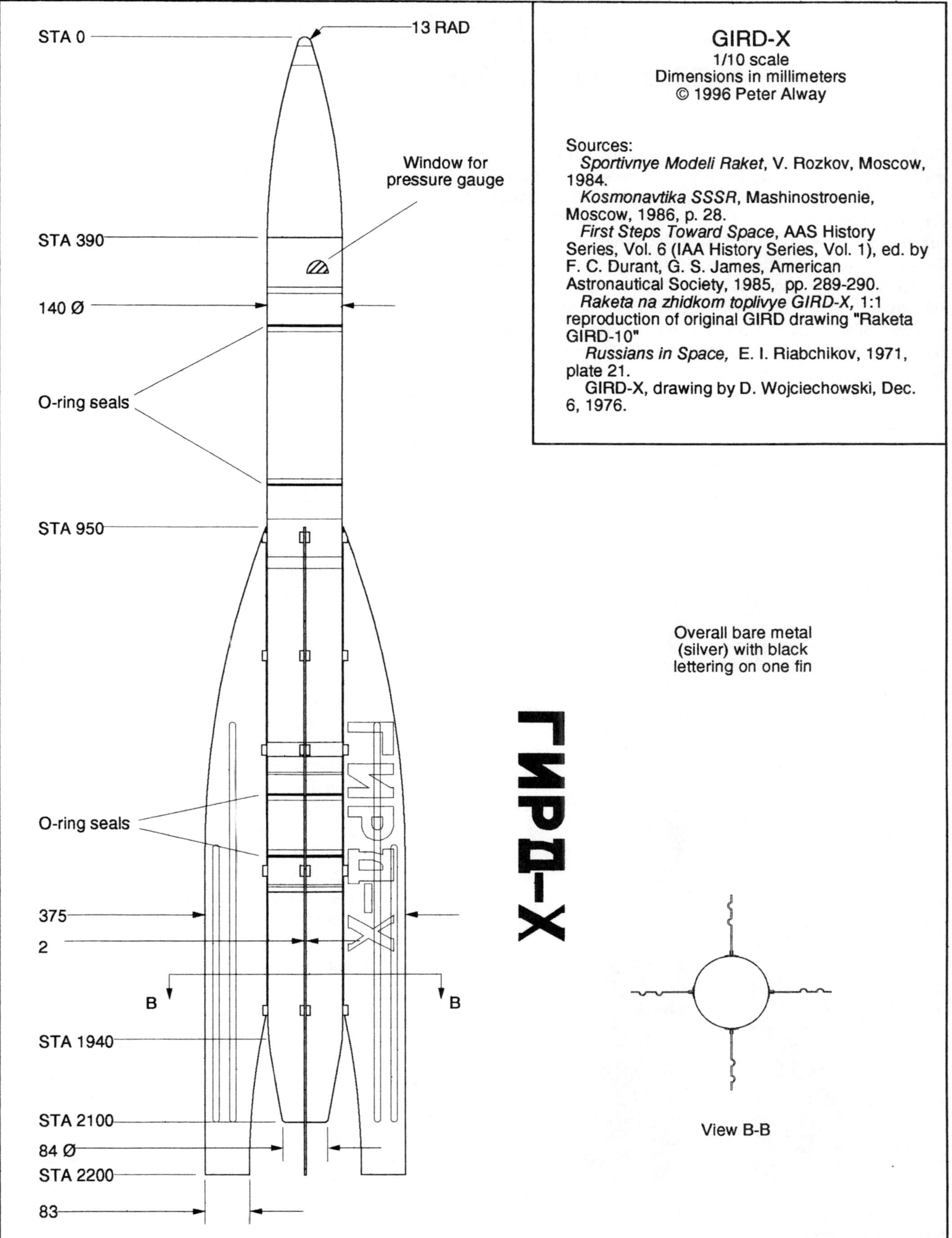

GIRD-X

1/10 scale
Dimensions in millimeters
© 1996 Peter Alway

Sources:
Sportivnye Modeli Raket, V. Rozkov, Moscow, 1984.
Kosmonavtika SSSR, Mashinostroenie, Moscow, 1986, p. 28.
First Steps Toward Space, AAS History Series, Vol. 6 (IAA History Series, Vol. 1), ed. by F. C. Durant, G. S. James, American Astronautical Society, 1985, pp. 289-290.
Raketa na zhidkom toplivye GIRD-X, 1:1 reproduction of original GIRD drawing "Raketa GIRD-10"
Russians in Space, E. I. Riabchikov, 1971, plate 21.
GIRD-X, drawing by D. Wojciechowski, Dec. 6, 1976.

Leningrad GIRD and the Razumov-Shtern LRD-D-1

The founding of Mos-GIRD (Moscow Group for the Study of Reaction Propulsion) led to the formation of groups elsewhere, most notably in Leningrad. Vladimir Vasilyevich Razumov, a fresh naval engineering graduate, founded Len-GIRD in 1931. Like Mos-GIRD, the organization was sponsored by Osoviakhim, a peculiarly Soviet organization for developing technical talent in government-supported hobbies and quasi-volunteer initiatives. Like Mos-GIRD, Len-GIRD embarked on a program of education, "technical propaganda," and rocket building. And like Mos-GIRD, Len-GIRD made do with facilities scrounged with the aid of any connections available to a group of technical enthusiasts. The group managed to find space to build rockets at the Institute for Wire Communication.

The Razumov-Shtern LRD-D-1 only flew in a solid-propelled test. (NASA History Office)

Among the solid propellant rockets planned was a photographic rocket designed to measure its own altitude through photos of a known baseline on the ground. While this project was never completed, by 1932, Len-GIRD was flying small solid propellant rockets to altitudes of about a kilometer. Designed by member M. V. Gazhala, these 2.1 meter tall, 23 cm-diameter rockets could do no more than disperse leaflets.

In 1932, Razumov conceived Len-GIRD's most ambitious rocket, the LRD-D-1. This was to carry meteorological instruments to an altitude of 5 km, and return them to Earth by parachute. Alexandr Nikolayevich Shtern, of Osoaviakhim's Air Equipment Design Bureau undertook the design of the LRD-D-1's liquid propellant engine. Shtern pursued a unique solution to the problem of forcing liquid oxygen and gasoline into the rocket's combustion chambers. His was a rotary engine, in which two combustion chambers spun rapidly around the rocket's vertical axis. Centrifugal force would feed propellants through radial pipes into the two chambers. Spin would be maintained by off-axis thrust from each nozzle. Each nozzle would be directed straight back, but the lower ends of the expansion nozzles would be cut diagonally. Not only would the engine's spin provide fuel pressure, it would also add stability. Stability was also enhanced by huge curving fins sweeping back from the rocket.

Work at Len-GIRD slowed in 1933 as members fell victim to a round of Stalin's purges, but in early 1934, Len-GIRD exhibited the rocket body and engine components at the All-Union Conference on the Study of the Stratosphere. Late in 1934, Len-GIRD tested the aerodynamics of the vehicle from the Aerolitic Institute station in Slutsk. As the liquid-propellant engine was not yet ready, this test was under the power of a solid rocket motor. By March of 1935, it became clear that Shtern's innovative design was unworkable, and the LRD-D-1 was abandoned.

Razumov-Shtern LRD-D-1 Specifications

Thrust	2000 N (440 lb)
Expected altitude	5 km (3 mi)
Total weight	90 kg (200 lb)
Structure weight	20.2 kg (44 lb)
Engine weight	16 kg (35 lb)
Gasoline weight	4.89 kg (11 lb)
LOX weight	17.5 kg (38 kg)
Length	2.565 m (8 ft 4.6 in)
Diameter	350 mm (13.8 in)

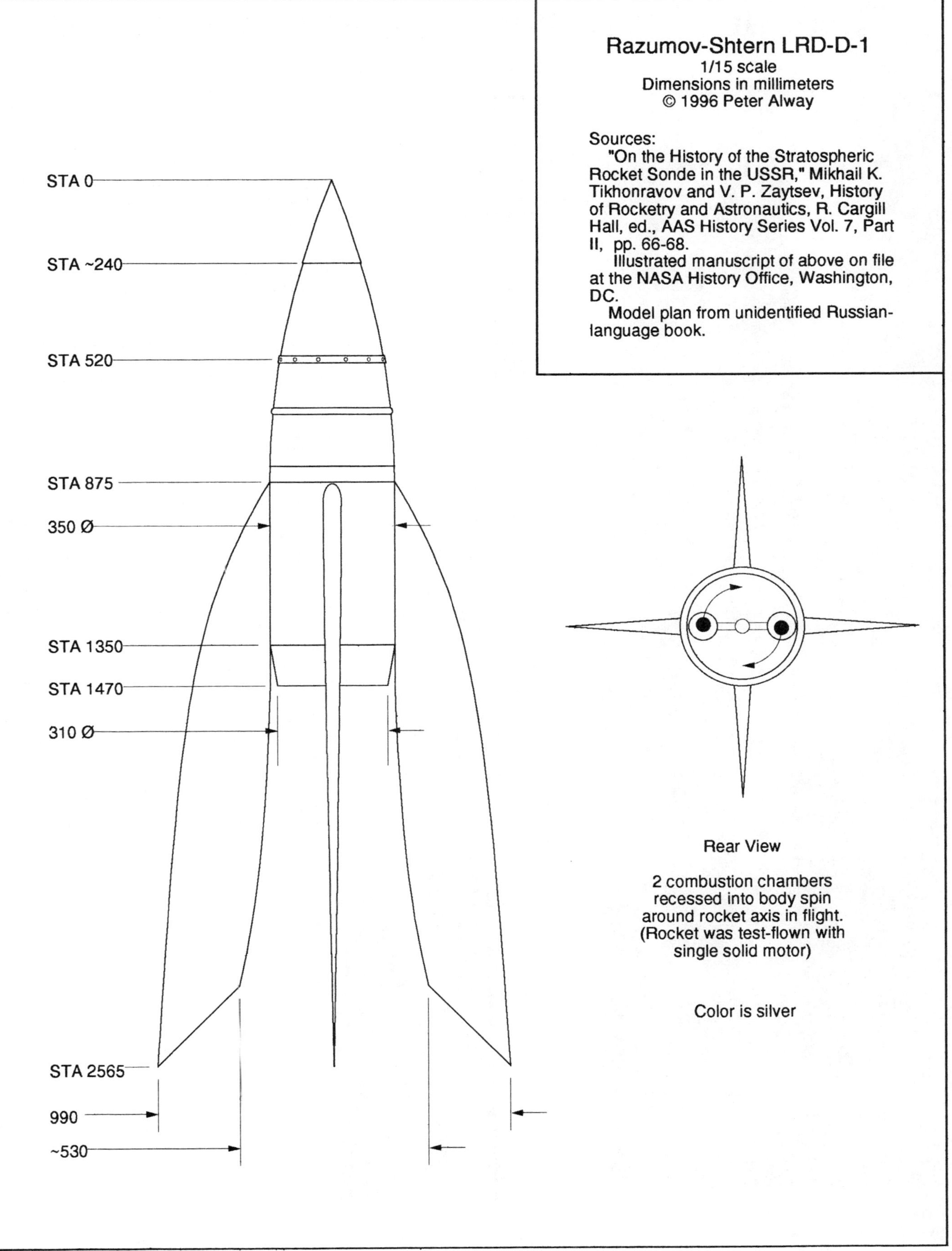

Razumov-Shtern LRD-D-1

1/15 scale
Dimensions in millimeters

Sources:
"On the History of the Stratospheric Rocket Sonde in the USSR," Mikhail K. Tikhonravov and V. P. Zaytsev, History of Rocketry and Astronautics, R. Cargill Hall, ed., AAS History Series Vol. 7, Part II, pp. 66-68.
Illustrated manuscript of above on file at the NASA History Office, Washington, DC.
Model plan from unidentified Russian-language book.

2 combustion chambers recessed into body spin around rocket axis in flight. (Rocket was test-flown with single solid motor)

Color is silver

RNII: Jet Propulsion Research Institute

Aviavnito on display in a Soviet Museum. (NASA History Office)

The successful launch of the GIRD-09 led to the creation of the Jet Propulsion Research Institute (RNII). GIRD's amateur rocket enthusiasts, most notably Sergei Korolev and Mikhail Tikhonravov, combined with the established Gas Dynamics Laboratory (GDL), led by Valentin Glushko, to create the new institute in September of 1933. RNII's efforts included static tests of advanced rocket engines, rocket-propelled gliders, "aerial torpedoes" (primitive cruise missiles), and ramjet vehicles. In later years, RNII would develop the solid-fuel Katyusha rockets of World War II. But RNII would also continue the vertical ascent rocket work of GIRD. The GIRD-X flew within months of RNII's formation in the fall of 1933.

GIRD-07

GIRD conceived its No. 07 rocket in 1933 as a test of liquid oxygen and Kerosene propellants. The rocket was distinguished by propellant tanks concealed within the fins. The pressure of evaporating oxygen forced the propellants into the engine at the front of the rocket. RNII finished the rocket, launching it with oxygen and alcohol on November 17, 1934.

Aviavnito

RNII's most successful vertical-launch vehicle was the Aviavnito rocket. The name is an abbreviation of the All-Union Aviation, Scientific, Engineering, and Technical Society. Aviavnito funded the rocket that bears its name to carry scientific instruments high in the atmosphere. RNII assembled the rocket in 1935 from an assortment of leftover components from various programs. The rocket body came from the GIRD-05 project, which was to burn kerosene with nitric acid. This vehicle was designed with four clustered tanks, faired to give it a squarish, rounded cross-section. The original GIRD-05 fins were discarded in favor of contoured hollow fins built for an RNII rocket designed by V. S. Zuyev. The engine was the 12-K, built by L. S. Dushkin. The ceramic-lined engine burned 90% alcohol with liquid oxygen. The launcher was a four-rail tower built for the GIRD-07.

The rocket carried a parachute and a barograph to measure altitude, which was predicted to be 5 kilometers (3 miles). Pravda covered the first launch on April 6, 1936:

> "The mechanic switched on the electrical primer. There was a grey cloud of evaporating oxidizer. A spark. Suddenly, a blinding yellow tongue of flame appeared at the bottom of the rocket. The rocket slowly moved upwards along the launcher guide rails, slipped their steel clutches, and hurtled swiftly upwards. The flight was an unusually effective and beautiful sight. A flame

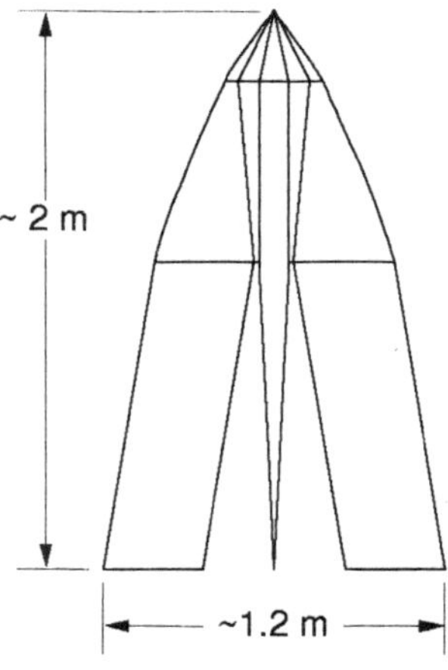

The GIRD-07 rocket. (Novosti photo at left, sketch by author at right)

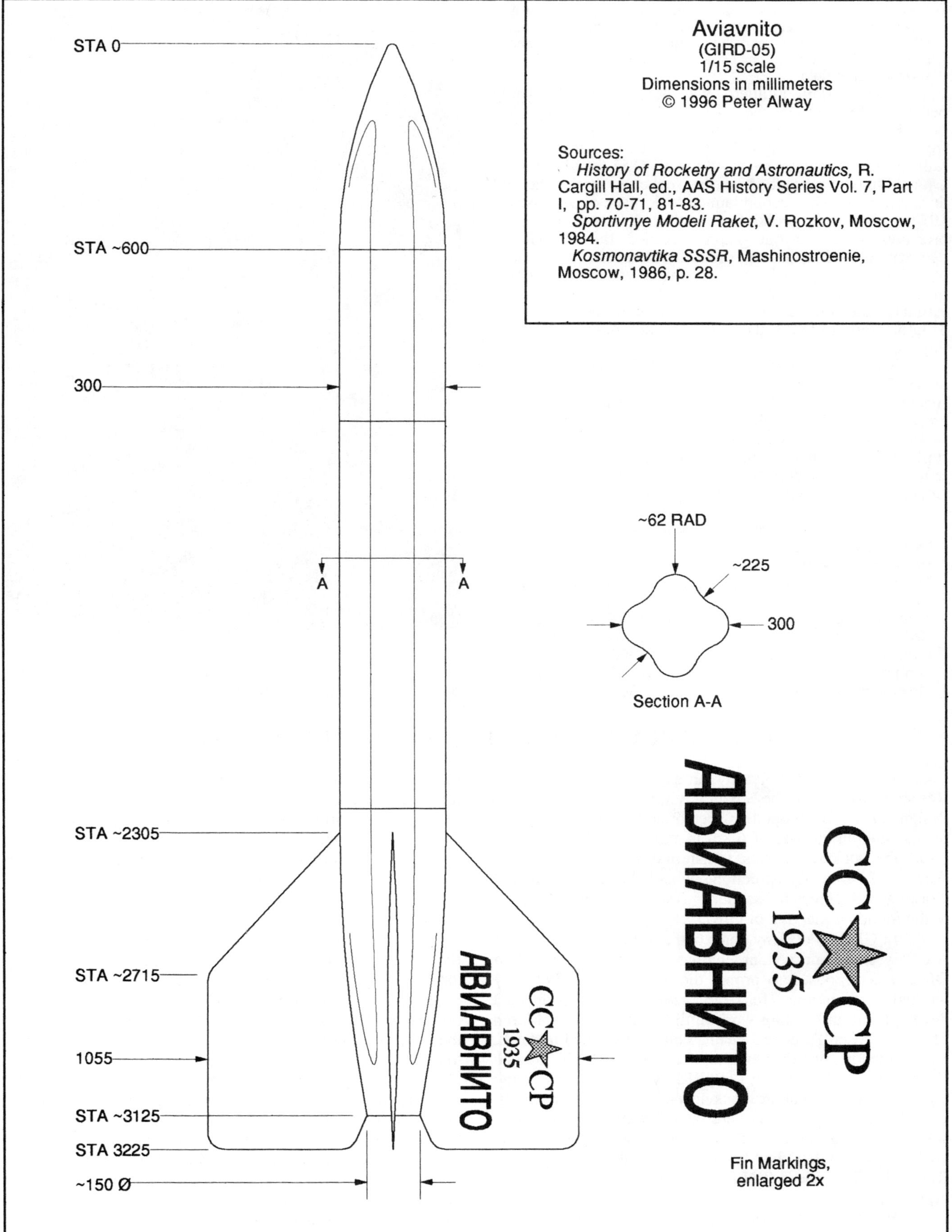
Aviavnito
(GIRD-05)
1/15 scale
Dimensions in millimeters
© 1996 Peter Alway
Sources:
History of Rocketry and Astronautics, R. Cargill Hall, ed., AAS History Series Vol. 7, Part I, pp. 70-71, 81-83.
Sportivnye Modeli Raket, V. Rozkov, Moscow, 1984.
Kosmonavtika SSSR, Mashinostroenie, Moscow, 1986, p. 28.
STA 0
STA ~600
300
A
A
~62 RAD
~225
300
Section A-A
STA ~2305
STA ~2715
1055
STA ~3125
STA 3225
~150 Ø
CC CP
1935
АВИАВНИТО
Fin Markings, enlarged 2x

flew from the motor's nozzle, and the gas efflux was accompanied by a deep hollow roar. After reaching a certain altitude, the white parachute in the rocket opened automatically, and the machine descended slowly to the snow-covered field."

The "certain altitude" was not revealed, probably because it fell short of the predicted 5 km. This may have been due to an inadequate launcher. The second flight flew from a rail launcher built from a stretch of narrow gauge railway rail attached to a 48-meter (157-ft)radio tower. Launch lugs were added to the rocket to grasp the rail during ascent. The second launch took place on August 15, 1937. The rail worked, but the parachute deployed prematurely. The barograph recorded the altitude of deployment as 2400 meters (8000 ft). Because of the speed of ascent, the parachute was ripped from the rocket instantly. The "calibrated eyeballs" of the launch crew estimated the vehicle rose to a maximum altitude exceeding 3 kilometers (10,000 ft). The rocket disintegrated on impact.

This was the last of the GIRD/RNII rockets to fly, as in 1937, Stalin's purges hit the rocketry program hard. Armaments Minister M. N. Tukhachevsky had overseen the creation of RNII, and allotted funds to rocket development. In 1937, he was framed as a Nazi spy, and many of RNII's staff were considered immediately suspect. Korolev was sent to a prison camp, as were untold others. The exact nature of RNII's disintegration has yet to come to light.

Aviavnito Specifications

Launch weight	100 kg (220 lb)
Propellant weight	32 kg (70 lb)
Thrust	3000 N (660 lb)
Duration	60 sec
Total Impulse	180,000 N-sec (40,000 lb-sec)
NAR designation	R 3000
Altitude	3000 m (10,000 ft)
Length	3.225 m (10 ft 7 in)
Maximum diameter	300 mm (11.8 in)

RNII's Aviavnito rocket attached to its guide rail for its second launch. (NASA History Office)

Korneyev's R-03 and Polyarny's R-06

In 1931, A. I. Polyarny, an engineer at the Scientific Research Institute of the Soviet Civil Air Force concocted a design for a solid propellant meteorological rocket capable of reaching an altitude of 6 km (4 miles). Before he could build the rocket, he was transferred to the Institute of Aircraft Engine Construction, where he found himself working with Fridrikh Tsander. Tsander invited Polyarny to the founding meeting of GIRD.

In GIRD, Polyarny contributed to the development to two key liquid propellant engines. Tsander's brainchild OR-2 was supposed to propel the RP-1 glider. Polyarny worked calculations for the project, and performed development and testing work until Tsander died in March of 1933. After Tsander's death, Polyarny worked under L. K. Korneyev on the engine for the GIRD-X rocket. Korneyev recalled his days at GIRD years later, describing how the quasi-amateur rocketeers, strapped for funds, contributed family silverware and jewelry to melt into solder for a rocket engine's cooling vanes. By the time GIRD-X flew, GIRD had become part of the Jet Propulsion Research Institute (RNII), a professional research organization.

By 1934, Polyarny and Korneyev were working independently of RNII on their own projects, designing a liquid propellant rocket that would never be built. In 1934 Osoaviakhim formed a jet propulsion section within its Stratospheric Committee. Polyarny took advantage of this new group to not just design a complete liquid-propelled rocket, but to build it. This "Osoaviakhim" or R-1 rocket, burning alcohol and liquid oxygen, would not fly for 3 years.

In 1935, Korneyev and Polyarny had their chance to fly rockets with their appointments as director and deputy director of the KB-7 design bureau. The exact nature of this organization—a deparment of RNII or an institution under military control—is not entirely clear. Their first task was the construction of a test station, including a large firing bay to test various liquid propellant rocket engines.

In 1936, Polyarny and Korneyev began rocket tests. Polyarny applied KB-7's automated ignition system to his Osoaviakhim rocket. The improved vehicle became the R-06. Korneyev had been working on his own alcohol-liquid oxygen rocket when KB-7 was created. Polyarny and his associates completed the design of Korneyev's vehicle, known as the R-03. KB-7 conducted static tests of both rockets inside the test station in late 1936.

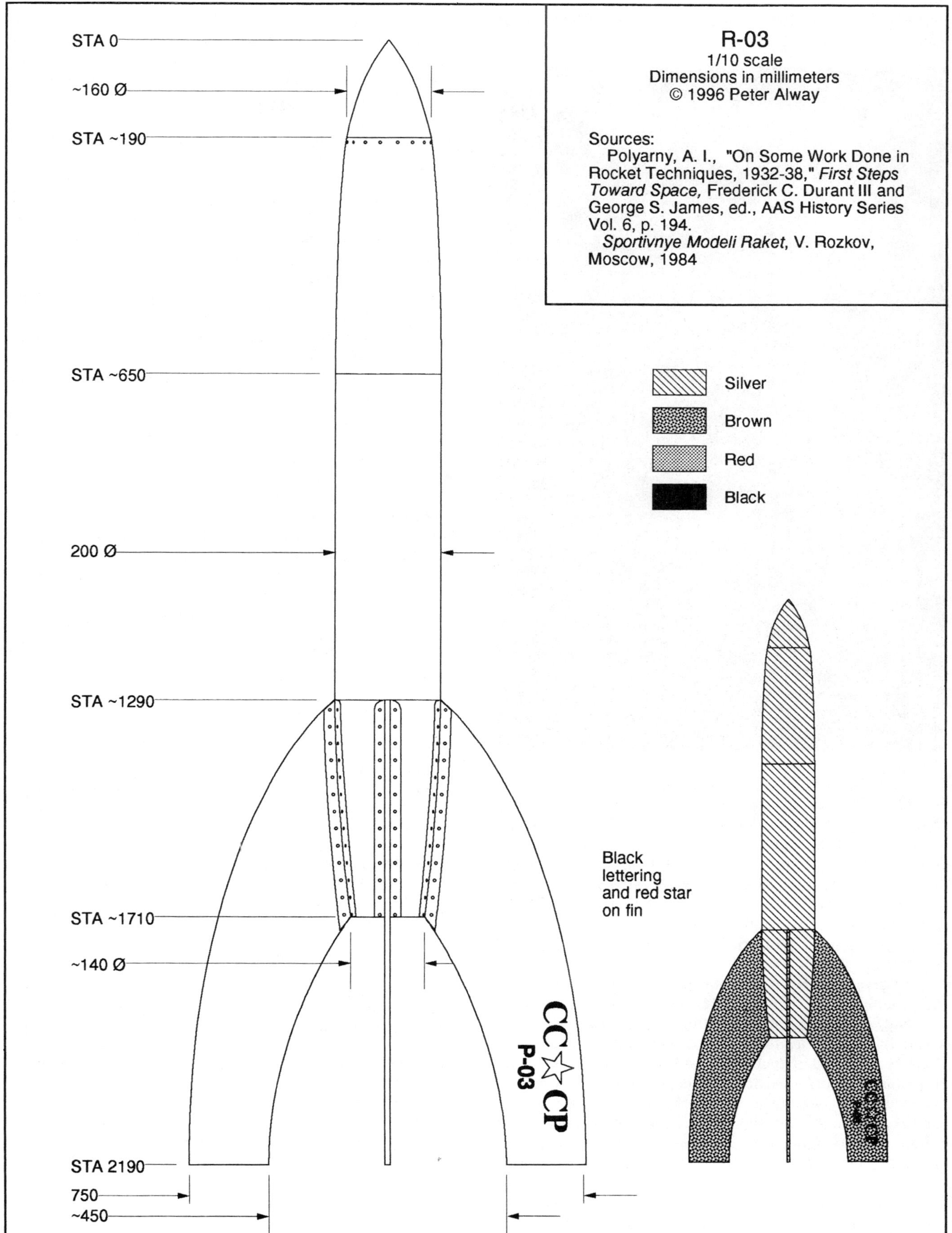
STA 0
~160 Ø
STA ~190
STA ~650
200 Ø
STA ~1290
STA ~1710
~140 Ø
STA 2190
750
~450
CC☆CP
P-03
R-03
1/10 scale
Dimensions in millimeters
© 1996 Peter Alway
Sources:
Polyarny, A. I., "On Some Work Done in Rocket Techniques, 1932-38," First Steps Toward Space, Frederick C. Durant III and George S. James, ed., AAS History Series Vol. 6, p. 194.
Sportivnye Modeli Raket, V. Rozkov, Moscow, 1984
Silver
Brown
Red
Black
Black lettering and red star on fin

Polyarny with his R-06 "Osoaviakhim" Rocket. (Smithsonian Institution photo no. 96-15699)

The first of ten R-03 rockets took to the skies in April of 1937. The last six flew in a campaign from February 25-28, 1938. While no altitudes are recorded, the rockets were launched at angles from 65°-75° from horizontal to ranges of 4 to 6.8 km (2 1/2 to 4 1/4 mi). These ranges imply altitudes as high as 3.7 km (2 1/4 mi)—far beyond the GIRD-09 and GIRD-X, and even out-performing RNII's Aviavnito rocket. The R-03 had no room for a payload or a recovery system; a pressurized air tank filled its elongated nose cone.

Polyarny's R-06 was the next to fly, on April 11, 1937. Unlike the R-03, the R-06 could carry a parachute in a cylindrical compartment at the base of its nose cone, which had space available for a meteorological payload. There is no record of such a payload being carried, however. Propellants were forced into the engine by the pressure of evaporating oxygen. Over the course of the 14 second burn, the pressure, and thus the thrust, dropped steadily. The design bureau launched a series of R-06 rockets in Leningrad on August 25, 1937 at a variety of elevation angles. The third rocket of the day, launched 70° above the horizon landed 6 km (4 mi) from the pad. As the R-06 carried a parachute, it is not clear what altitude this flight may have reached, but assuming a ballistic trajectory, one can calculate that this rocket may have reached an altitude exceeding 4 km (2 1/2 mi). The last two of nine R-06 rockets flew in late February of 1938. The final rocket, launched February 28, landed 3.1 km (1.9 mi) from the pad.

Korneyev and Polyarny's design bureau went beyond propulsion to explore gyroscopic guidance. The ANIR-5 featured a gyro fixed to the airframe between the oxygen and alcohol tanks of a modified R-06. KB-7 built six of these stub finned-rockets in 1938. The design was tested, but it is not clear if these were static or flight tests. The design bureau concluded that the fixed gyroscope system would not scale up to larger rockets.

KB-7 designed a remarkable assortment of rockets to explore the upper atmosphere, including the R-05, whose liquid-propellant sustainer was augmented by solid boosters, and the R-10, with two liquid stages, capable of reaching an altitude of 100 km (60 miles).

The story of Polyarny, Korneyev, and Design Bureau Number Seven ends with unfinished projects. V. P. Glushko reported that KB-7 was dissolved in 1938 or 1939 due to a failure to meet objectives. But this is almost certainly not the whole story. Soviet rocket history descended into a dark age in 1937-38, with the imprisonment of rocket designers in Stalin's purges. Whatever hardships may have befallen Korneyev and Polyarny, they both survived them. Yaroslav Golanov's 1973 biography of Sergei Korolev describes Korneyev as "now a senior engineer." Polyarny presented memoirs of his 1930's work at a 1967 space history conference. While the exact contribution of KB-7 to Soviet rocketry is not clear, the R-03 and R-06 were the highest-flying Soviet rockets of the 1930's. They may even have been the highest flying rockets in the world until the tests of the German A-5 in 1938 and 1939.

R-03 Specifications

Launch weight	34 kg (75 lb)
Fuel weight	12.5 kg (25.2 lb)
Thrust	1180 N (265 lb)
Duration	21 sec
Total impulse	25,000 N-s (5500 lb-s)
NAR designation	O 1180
Expected range	8.5 km (5.3 mi)
Length	2.6 m (8 ft 6.4 in)
Diameter	200 mm (7.9 in)

R-06 Specifications

Launch weight	10 kg (22 lb)
Fuel weight	2.4 kg (5.3 lb)
Payload weight	0.5 kg (1.1 lb)
Thrust	245-390 N (55-88 lb)
Duration	14 sec
Total impulse	4400 N-s (1000 lb-s)
NAR designation	L 245
Expected altitude	5 km (3 mi)
Length	1.7 m (5 ft 6.9 in)
Diameter	126 mm (5.0 in)

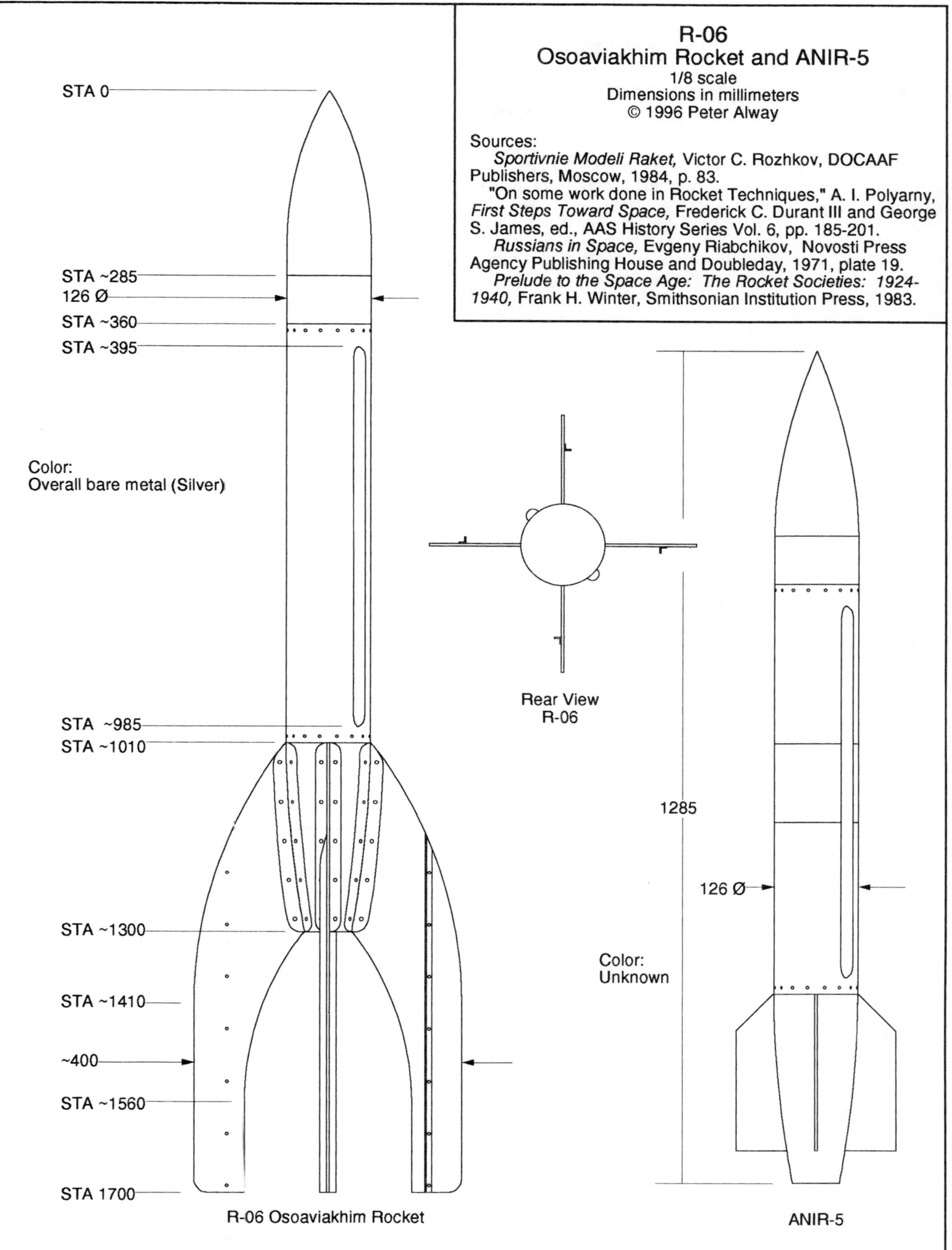

R-06 Osoaviakhim Rocket

ANIR-5

Building Flying Models of Early Rockets

An Introduction to Model Rocketry

The rocket experimenters of the early 20th century suffered a terrible rate of injury and death from their pyrotechnic experiments. Clarence Hickman blew off several fingers while assisting Goddard with solid fuel work. Max Valier died when a liquid-propelled engine exploded, sending shrapnel through his aorta and killing him in minutes. Reinhold Tiling and two assistants died of burns from an explosion and fire while pressing gunpowder for a rocket. A bystander at an ARS static test suffered a broken elbow in an explosion. Robert Esnault-Pelterie was forced to stop experimenting when he blew off much of a hand experimenting with high-energy propellants.

In the late 1950's, space travel was all the rage in the United States, and young amateur rocket experimenters began to repeat the carnage on themselves. In response, Nebraska pyrotechnician Orville Carlisle and New Mexico rocket professional G. Harry Stine created the hobby of model rockety, in which the pyrotechnic work is left to professionals with automated equipment. For the first time, hobbyists could engage in safe rocketry.

The heart of model rocketry is the model rocket engine. These small brown or black cylinders, ignited electrically from a distance, provide all the thrust a rocket needs to reach an altitude of hundreds or even thousands of feet. Once at altitude, they provide an ejection charge—a small kick of hot gas—to deploy a recovery system. After the flight, only the spent engine is discarded, leaving the rocket ready for another flight. High power rocketry is a similar hobby, subject to additional regulation, but allowing the use of larger engines and heavier models.

By following the model rocket and high power rocket safety codes, modelers can recreate the rockets of the 1930's and their flights, without recreating the tragic accidents. Many of the rockets described in this book could even be modeled at full scale and flown to their original altitudes within the bounds of high power rocketry.

Modelers interested in both model and high power rocketry are advised to join the National Association of Rocketry (NAR—an application is on the last page of this book). The NAR publishes *Sport Rocketry* magazine, supports sections around the country, and sanctions contests and sport launches. Aficionados of more powerful rockets may wish to join the Tripoli Rocketry Association (TRA). TRA members receive *High Power Rocketry* magazine and can join local clubs around the US. Both organizations offer certification for use of high power hobby rockets.

Model Rocket Construction Techniques

Before building from the plans in this book, the modeler should build a few basic parachute-recovered model rocket kits from Estes, Quest, or one of their competitors. Many of these models are not especially difficult, and they will acquaint the hobbyist with how the basic parts of a model rocket—body tube, fins, nose cone, engine mount, and recovery system fit and work together. In fact, the kits themselves are excellent sources of parts. There are a few important techniques in these plans that you won't find in kit instructions. Most are fairly simple. Only turning a nose cone is especially challenging. Fortunately, there is a supplier of custom balsa nose cones that can eliminate this step at a reasonable cost.

The techniques you will need to complete the plans in this book are discussed here. For more techniques in scale modeling, refer to *The Art of Scale Model Rocketry,* available from Saturn Press (1994). For a general overview of Model Rocketry, refer to *The Handbook of Model Rocketry,* 6th edition, by G. Harry Stine (Wiley, 1994).

Cutting Tubes

Mark the length of tubing desired. Tear a strip from a sheet of paper. Wrap it tightly around the tube a couple times to insure the factory-cut edge is truly perpendicular to the tube. Tape the paper strip to hold its alignment (if you have a piece of tubing that slips over the tube you are

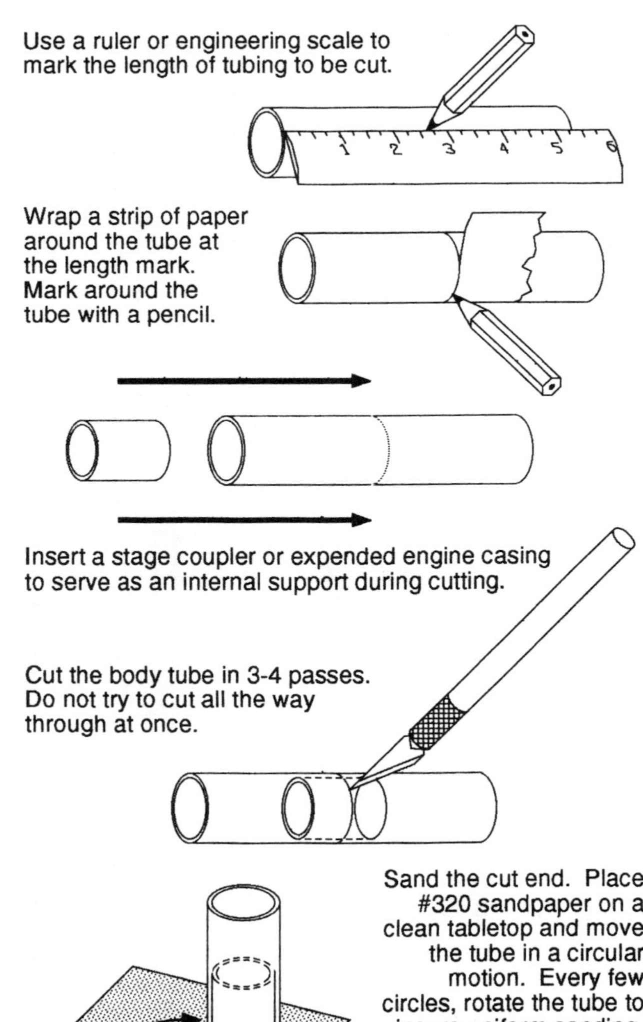

cutting, use it instead of paper). Draw along the factory-cut edge with a pencil (you can use a pen if that part of the tube will be painted a dark color or metallic silver).

Give the tube some internal support while you cut. A stage coupler or expended engine under the cut works well.

Cut with light, even pressure while rotating the tube. It will take three or four turns to cut through the tube.

To sand the cut, slide the internal support (stage coupler or engine) flush with the cut end of the tube. Lay sandpaper on a smooth tabletop and hold the tube vertically. Sand the whole cut with a circular motion. Rotate the tube periodically to insure even sanding around the cut.

Cutting Fins

You will need to cut fins from wood for these models. Balsa is extremely lightweight, but it has a very open grain and a soft surface. Basswood, available in good hobby shops, is much easier to fill for painting. Model aircraft plywood is useful for parts that need its bi-directional strength, including the odd-shaped fins of many 1930's rockets.

You may want to substitute plywood fins in the GIRD-09 plans. If you do, add through-the-wall tabs or thick epoxy fillets—glue does not bond well to the root edge of a plywood fin.

Begin by photocopying from the plan. Cut out the pattern, and trace the outline onto the wood. Pay attention to the direction of the wood grain, as this must be oriented correctly for maximum strength.

Cut basswood or balsa with a hobby knife with a fresh blade. Don't try to cut through basswood or plywood in a single pass. Once the fins are cut out, hold them together in a stack and sand the edges on a table top. Check your progress against fin pattern (cut exactly to size). Use an emery board to shape root edges to match boattails exactly.

Some of the fins of the original rockets in these plans were simple flat sheet metal, with no airfoil. If you wish to round the edges or airfoil your fins, be sure to leave the root edge flat.

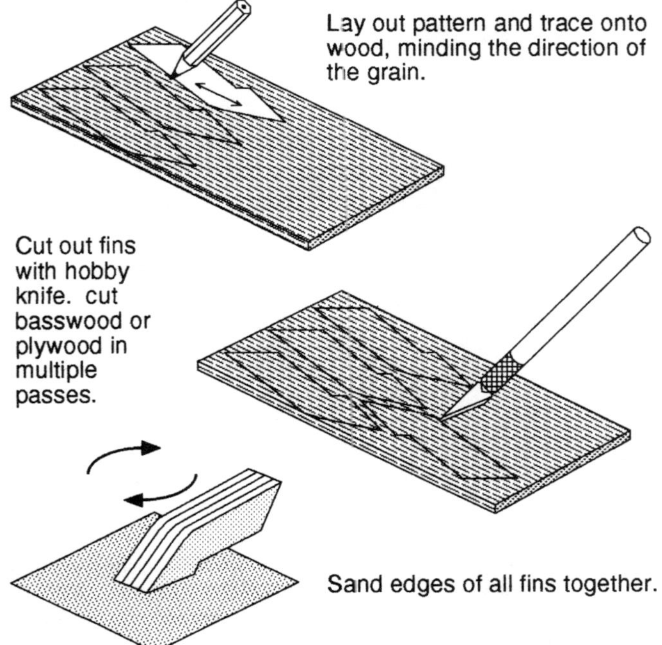

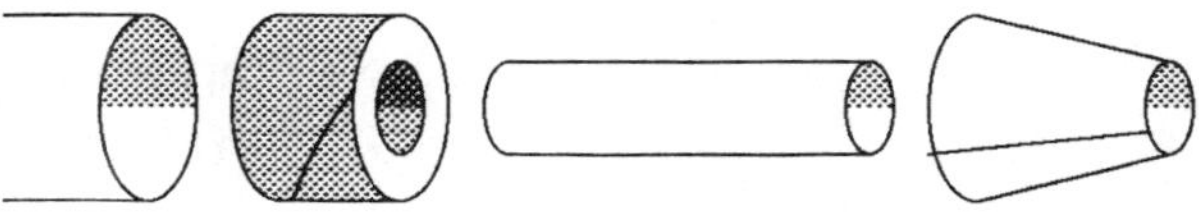

Assembly of paper adapter section

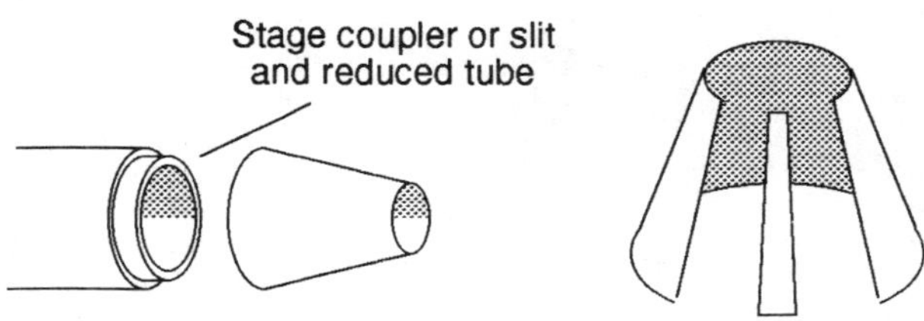

Alternative support for large end of shroud: An inside tab produces a less conspicuous seam than an overlap tab

Paper Transitions

In principle, paper transitions, such as the boattail of the Goddard A-series rocket, are very simple. Photocopy them onto light cardstock (available at most copy shops). When in doubt, copy onto different weights of stock. It is also wise to make extras. Cut the copy out, glue a tab, and it's done. You can eliminate the tab overlap if you cut off the tab, and glue it to the inside of the joint.

After cutting out a shroud, it often helps to put some curl into the paper before gluing it. Rub the paper against a dowel or other cylinder to avoid creasing the paper. Water-based glues such as Elmer's or Titebond can warp or wrinkle thin paper. Glues such as Duco cement or Testors wood glue are ideal for gluing thin paper. Hold the seam over a mandrel such as a dowel or small body tube while the glue dries to give the seam some curvature.

The large end of a paper transition must rest on the outside of a tube. This tube is usually a stage coupler that fits into the main tube at the large end of the transition. The order of assembly is usually to glue adapter rings to the stage coupler, test fit the coupler-ring assembly over the small tube (the engine tube for a conical boattail), and test fit the conical shroud over the small tube and the stage coupler. Then slide the stage coupler-ring assembly into the correct position, and glue it into place. Next glue the conical paper shroud in place, and finally glue the whole assembly into the larger tube.

There is a special case in which you can skip tapered adapters altogether. The plastic nose cone of the Estes Black Brant II kit has the correct taper for Goddard's L-10 through L-30 rockets. Cut off the tip, shortening the cone to about 4" to build 1/6.8 scale models.

Custom-Turned Balsa Nose Cones

Balsa Machining Service (BMS) offers custom-turned transitions as well as conical and ogive nose cones for modelers without the equipment for wood turning. BMS will produce tangent ogive and conical nose cones up to 3.937" in diameter and 10" long. The Goddard A-3 plans in this book give the dimensions BMS requires to produce an accurate nose cone. Contact BMS at 1002 Florence Street, Lemont, IL 60439, USA (phone (708) 257-5420, fax (708) 257-0341.

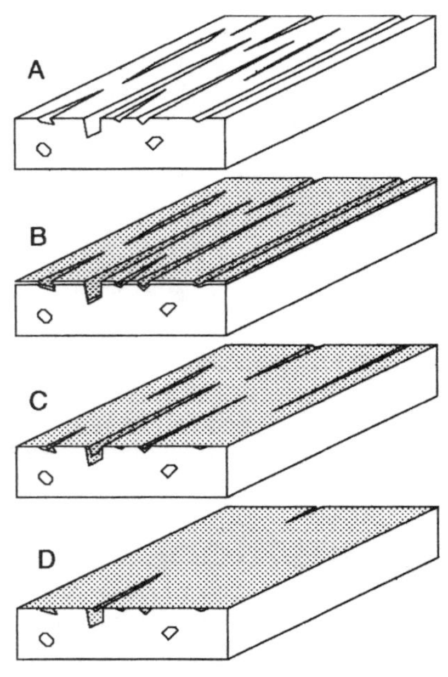

Sanding and Sealing Balsa.

A: Open grain (magnified) of balsa must be filled.
B: Layer of glue or sealer.
C: After sanding, only deeper depressions remain.
D: Another cycle of sanding and sealing fills most grain.

A few more cycles will give the balsa a perfect finish.

Basswood may be ready to paint after just one or two cycles.

Finish Preparation

The metallic paints required for models of early rockets amplify every flaw in a model's surface. It is critical to fill in both the grain of wood parts and the spiral grooves of cardboard tubing.

Begin by sanding all wood surfaces with #320 grit sandpaper. I have used four materials for sanding sealer. Elmer's glue and Titebond are easy to find, and there are times when they are a superior material. They strengthen balsa fins and make it possible to glue parts to the sealed surface. Sig sanding sealer and Hobby Poxy Fast Fill tend to dry a bit faster and sand more easily.

I usually begin with a coat of Titebond smeared onto the wood with the straight edge of a scrap of wood or an unwanted credit card. Press down while spreading the glue to force it into the wood and to simultaneously remove all but the thinnest possible layer of glue from the surface. Be sure to coat the edges. For nose cones, just use your fingers.

Once the glue is dry, sand it with #400 grit sandpaper. My preference is to apply one or two coats of glue before moving on to sealer. If you don't have sanding sealer on hand, you can repeat the process with glue until the wood is smooth.

Both Sig sanding sealer and Hobby Poxy Fast Fill work well for additional coats (some modelers prefer to apply sealer or thick coats of spray primer after the model is assembled). The opaque white Fast Fill stands out against the wood and glue during sanding. It can be brushed on or spread on, although brushing is preferred. Be sure not to apply sanding sealer where you will be gluing later (root edge of a fin). Sand each coat of sealer down to the level of the previous layer. The contrast between the white Fast Fill and the darker transparent Titebond surface makes it easy to see the progress of sanding. One or two coats works for bass wood, while balsa requires three or four. You might want to use a coat of white spray paint as a final layer of sealer and as a test of the surface.

Body tube seams also require filling. I've used Hobby Poxy Fast Fill for this, with reasonable results. Others use Hot Stuff and microballoons, spackling compound, and other filling materials. Use a fine paint brush to fill the spiral seams with your choice of filler. Carefully scrape away the excess. Then touch up the places where you completely scraped the filler. Scrape your touch-up. You may want to proceed with construction at this point. To complete sealing of the tube seams, spray with a couple thin coats of white paint (I use Dutch Boy Fresh Look flat white; some modelers prefer Krylon white sandable primer). Look for remaining gaps along the seam, and fill. Lightly sand with #400 grit finishing sandpaper. Spray... sand... spray... sand... Keep this up for three to four cycles, and you will not only fill the spiral seam itself, but also the more subtle spiral contours on the body tube. You'll have to cut away the paint where you want to glue the fins on.

Painting

It is difficult to get a convincing silver finish on a model—plastic modelers have a plethora of products for giving convincing metallic finishes. Bear in mind that a flying model will be subject to more abuse than a plastic model, and the more fragile finishes may not be suitable for flying models.

Testors silver spray paint is one place to start. The matte aluminum finish of this paint may not be the most convincingly metallic, but it is inexpensive and dries quickly. Contrasted with a glossy silver, such as Dutch Boy/Fresh look, you can produce a convincing impression of a two-metal finish. Testors also manufactures "Metalizer" paints that can be buffed to a high gloss.

One problem with metallic paints is that their surfaces are easily damaged by tape used in masking. Sometimes it is best to paint separate sections of a body tube with different paints and join them with stage couplers (these can be improvised by slitting scrap bits of tubing). An alternative is to use self-adhesive chrome Monocote trim to contrast with silver paint. Some modelers tone down its aluminized finish with steel wool, giving yet another finish.

The four plans in this book just suggest a few possibilities. It is my hope that one of the other rockets in this book grab your imagination, and that you build it to whatever scale you please, even full size.

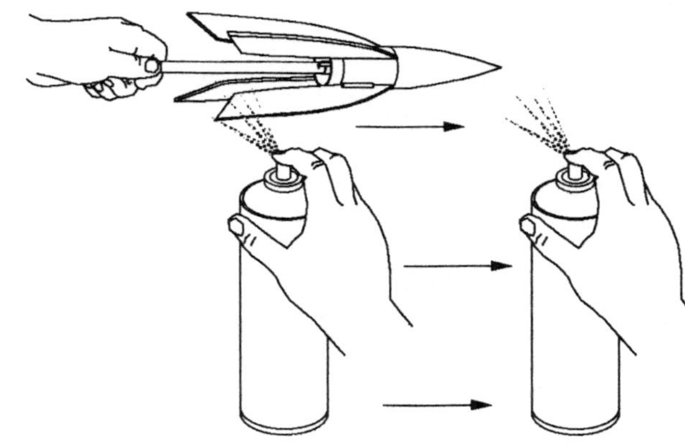

Hold the model by a 1/2" dowel while spraying in long strokes running past both ends of the model. Add an expended engine to the end of the dowel to fit standard-engine models.

Goddard A-3

1/5.5 Scale

Parts list:

A: Nose cone, hand turned from balsa, or custom ordered from Balsa Machining Service
B: Parachute, 18"
C: Shock cord and mount
D: Tube extension, 2 1/2" Estes BT-60
E: Stage couplers (2), Estes JT-60 from large tube coupler pack
F: Body tube, 18" Estes BT-60
G: Launch lug
H: Adapter rings AT-2060
I: Engine mount tube: 8 1/2" Estes BT-20
J: Engine block, cut from end of expended standard engine
K: Tail shroud, cut from cardstock photocopy of pattern
L: Fins (4) from 1/16" basswood or balsa
M: Screw eye

Notes on tail section assembly:
First glue adapter rings, H, to either end of one stage coupler and allow to dry. Then curl and glue tail shroud, K, and allow to dry. Test fit ring-coupler assembly over engine tube , I, and then fit shroud , K, over engine tube and ring-coupler assembly. Adjust for snug fit of shroud over ring-coupler assembly, with 1/16" of tubing protruding from bottom end of shroud. Glue rings to engine mount tube, and when dry, glue tail shroud in place. Finally glue completed tail section into the main body tube, F.

Note: If separate parts are not available locally, you may wish to use an Estes Big Bertha kit for the engine mount tube and rings, and an Estes Mean Machine for stage couplers and extra body tube. Leftover parts can be used in the GIRD-09 or HW-II models described in this book.

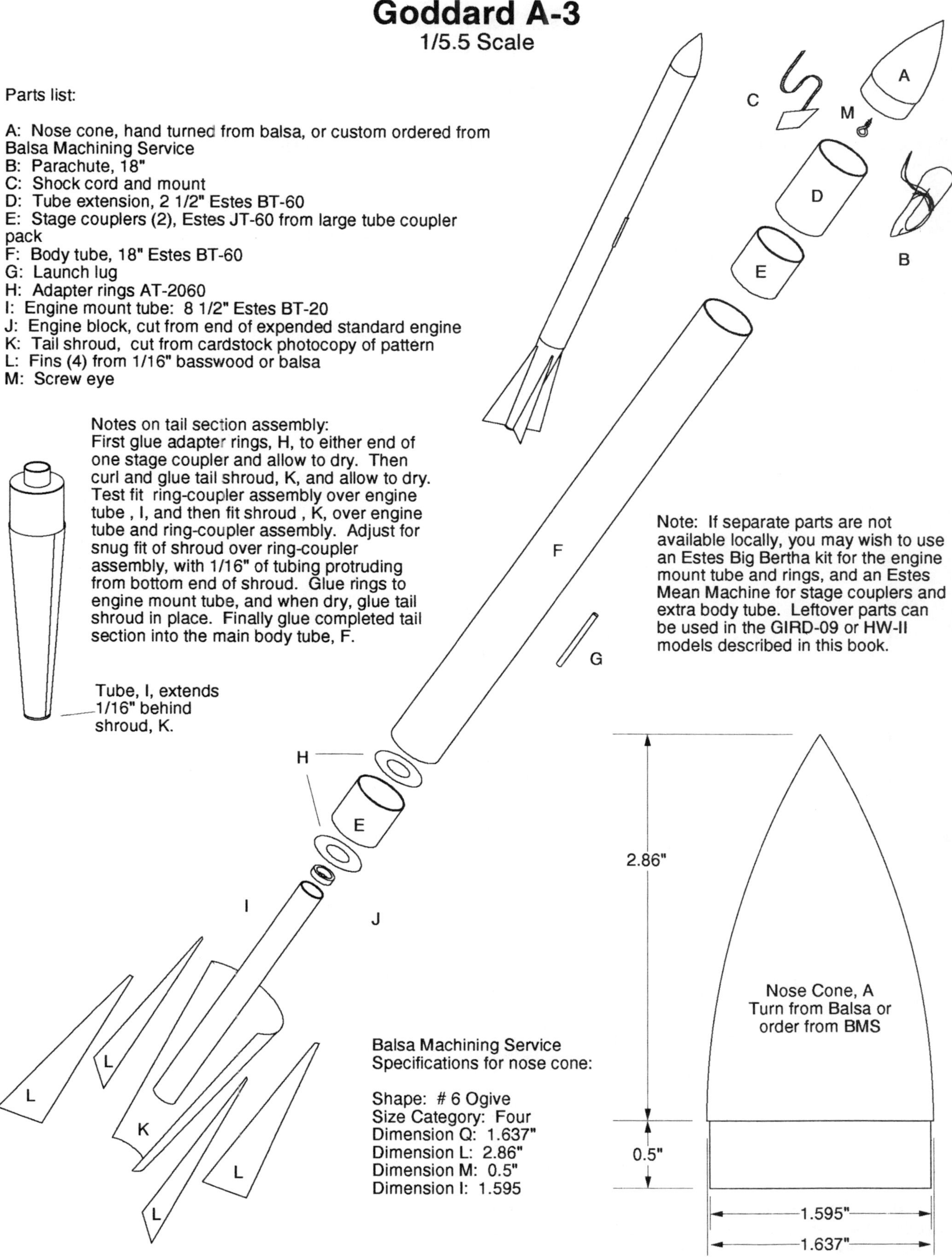

Balsa Machining Service
Specifications for nose cone:

Shape: # 6 Ogive
Size Category: Four
Dimension Q: 1.637"
Dimension L: 2.86"
Dimension M: 0.5"
Dimension I: 1.595

Goddard A-3
1/5.5 Scale
Patterns

Photocopy this page onto cardstock. Make several copies for practice, as the long tapered section can be difficult to glue.

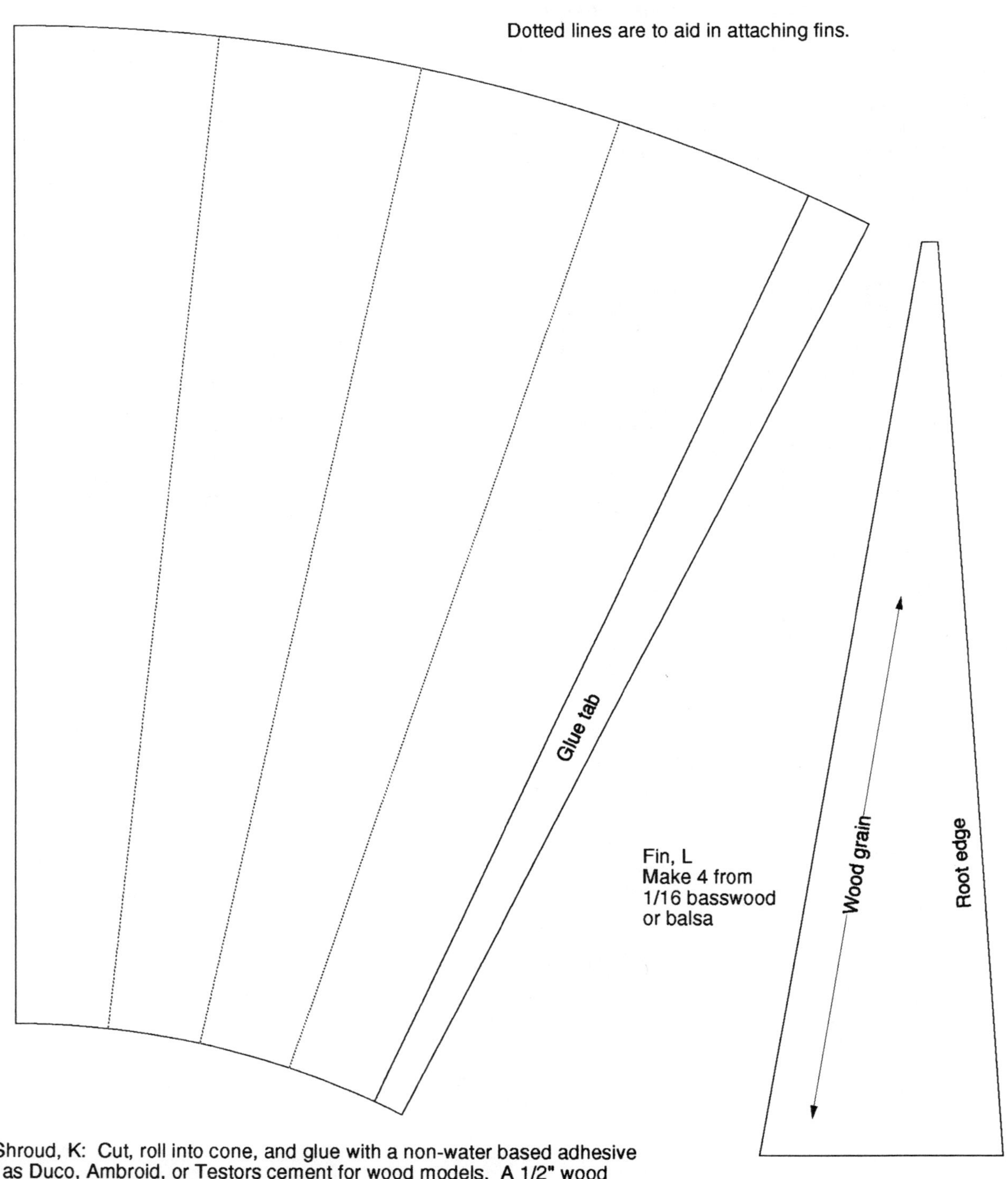

Tail Shroud, K: Cut, roll into cone, and glue with a non-water based adhesive such as Duco, Ambroid, or Testors cement for wood models. A 1/2" wood dowel running through the cone can aid in the gluing process.

Winkler HW-II

1/9.5 Scale

Parts list: (parts available from Estes Industries)

A: Nose cone, PNC 60 AB (from Big Bertha), shoulder shortened to 1/4"
B: Parachute, 12"
C: Shock cord and mounts
D: Launch lugs (2) 1/2" long
E: Adapter ring RA-555
F: Engine mount tube: 3" BT-5
G: Engine block, cut from end of expended mini engine
H: Tail section, PNC-60AH (from Mean Machine or NC-60B set), ends cut off as shown
I: Fins (3) from 1/16" basswood or balsa

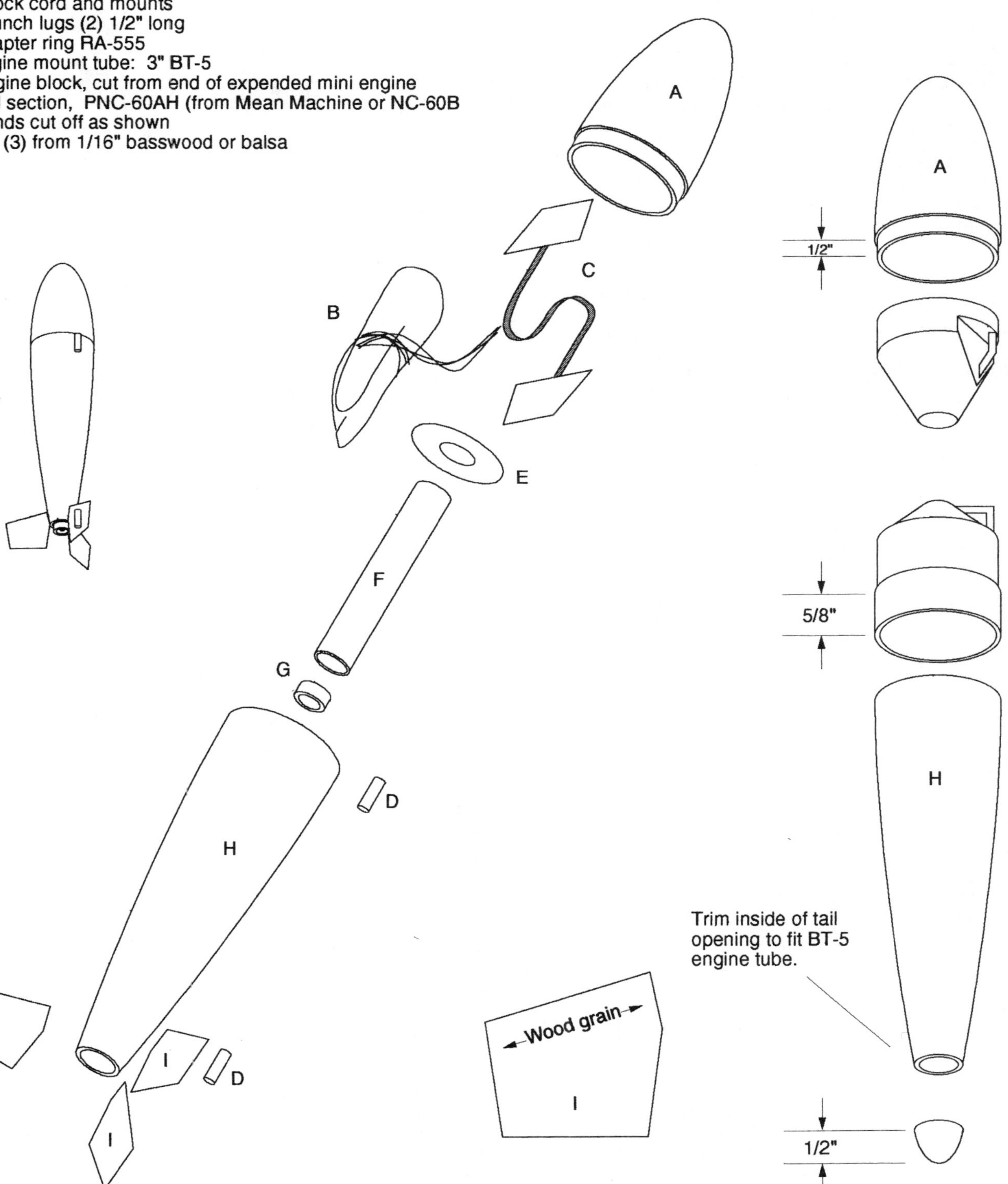

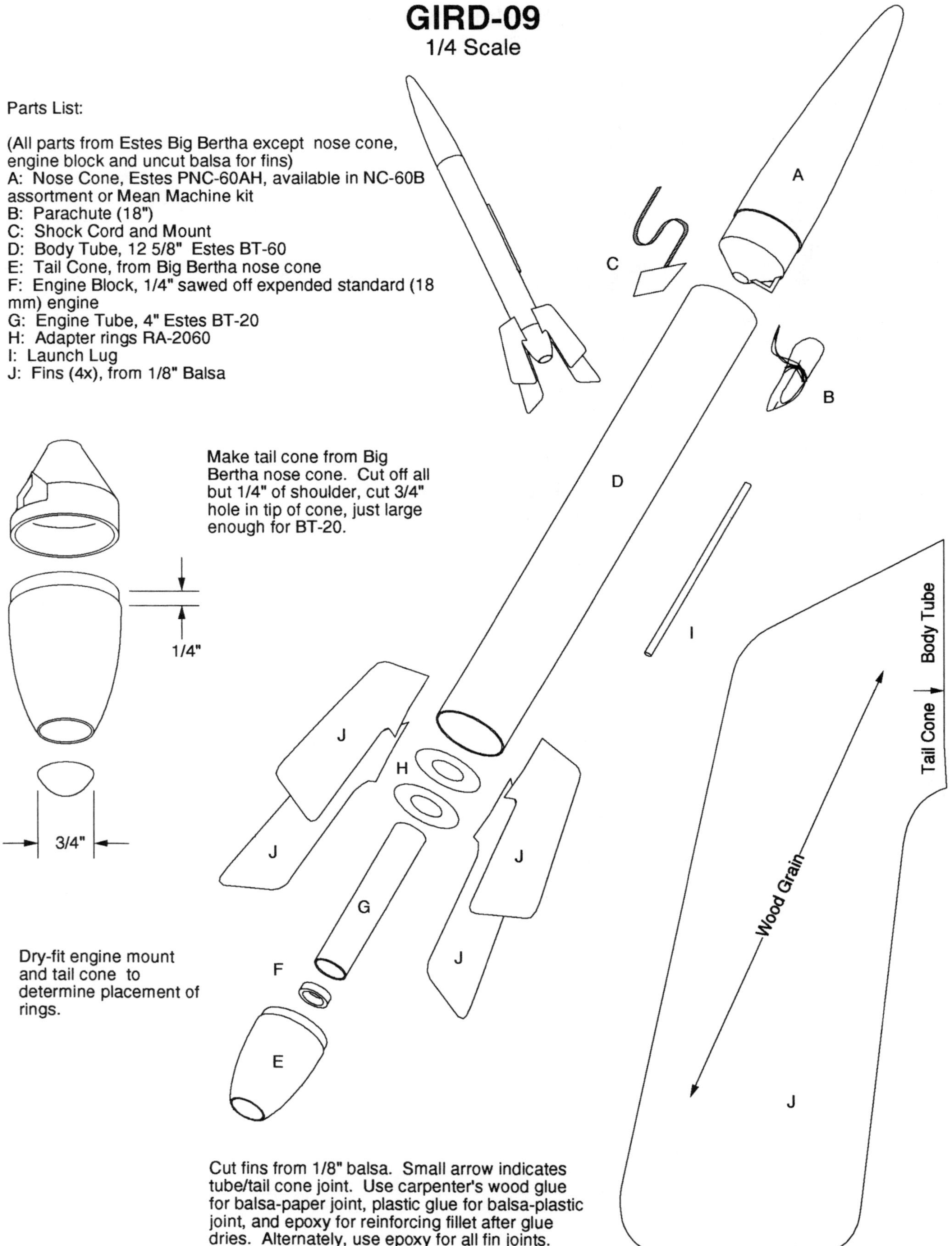
GIRD-09
1/4 Scale
Parts List:
(All parts from Estes Big Bertha except nose cone, engine block and uncut balsa for fins)
A: Nose Cone, Estes PNC-60AH, available in NC-60B assortment or Mean Machine kit
B: Parachute (18")
C: Shock Cord and Mount
D: Body Tube, 12 5/8" Estes BT-60
E: Tail Cone, from Big Bertha nose cone
F: Engine Block, 1/4" sawed off expended standard (18 mm) engine
G: Engine Tube, 4" Estes BT-20
H: Adapter rings RA-2060
I: Launch Lug
J: Fins (4x), from 1/8" Balsa
A
B
C
D
Make tail cone from Big Bertha nose cone. Cut off all but 1/4" of shoulder, cut 3/4" hole in tip of cone, just large enough for BT-20.
1/4"
3/4"
I
Body Tube
Tail Cone
Wood Grain
J
J
H
J
J
G
J
F
E
Dry-fit engine mount and tail cone to determine placement of rings.
J
Cut fins from 1/8" balsa. Small arrow indicates tube/tail cone joint. Use carpenter's wood glue for balsa-paper joint, plastic glue for balsa-plastic joint, and epoxy for reinforcing fillet after glue dries. Alternately, use epoxy for all fin joints.

Razumov-Shtern LRD-D-1

1/8.4 Scale

Parts List (Estes Parts)

A: Nose Cone for BT-60, from payload section PS-5560 kit, shoulder cut to 1/4" long.
B: Body Tube 2 1/4" BT-60
C: Plastic BT-55/BT-60 adapter, from payload section PS-5560 kit, taper cut 1/2" from BT-60 shoulder, BT-60 shoulder cut to 1/4"
D: Engine Mount tube, 1 3/4" BT-5
E: Adapter Rings (2) RA-560 from multi-purpose adapter set, or cut from cardboard (0.541" ID, 1.595" OD)
F: Mini Engine hook
G: Fins (4), from 1/8" balsa
H: Launch lug, 1 1/4" long, for 1/8" rod
I: Shock cord, 1/8" x 18"
J: Parachute PK-12

Cut a tiny hole in the upper adapter ring (E), tie cord (I) around engine mount tube (D), and run cord up through hole.

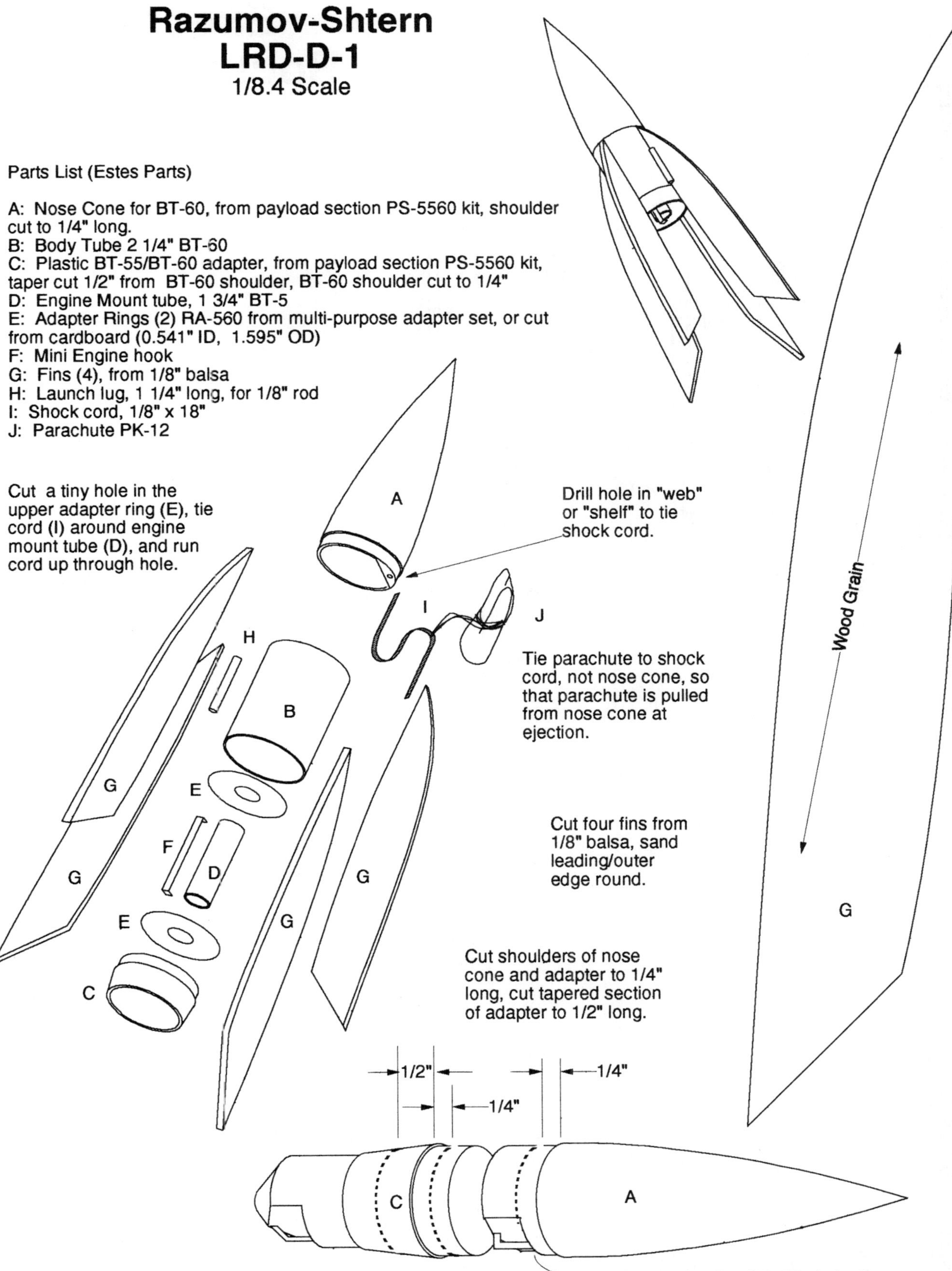

Experimental Rockets to 1941

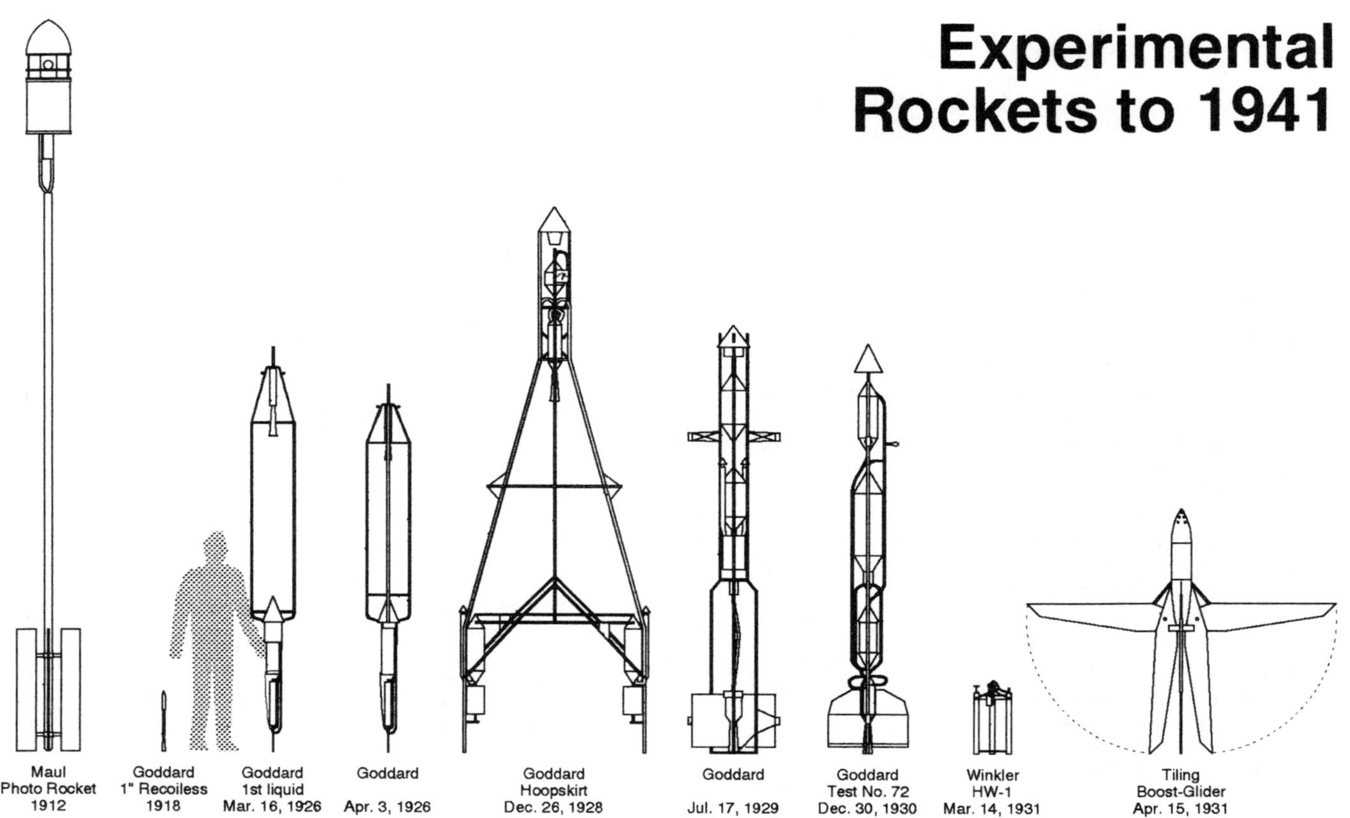

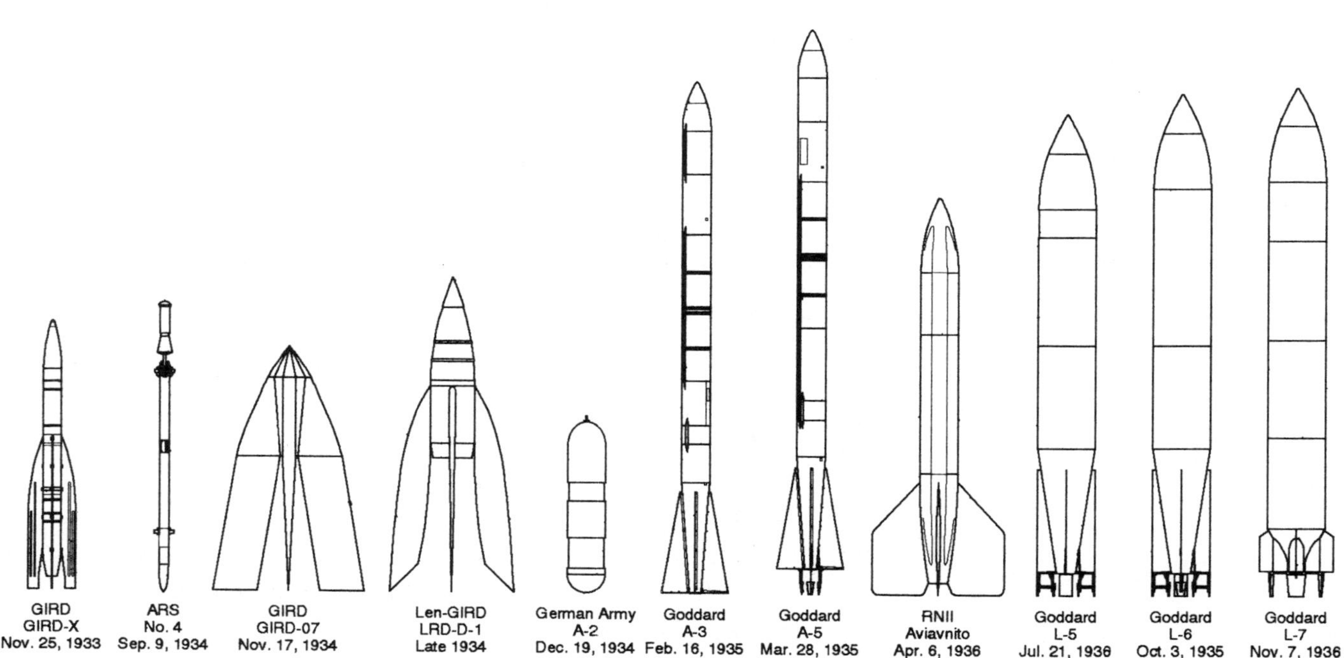

The rockets appearing in *Retro Rockets: Experimental Rockets 1926-1941* are shown here at a constant 1/60 scale, in chronological order.

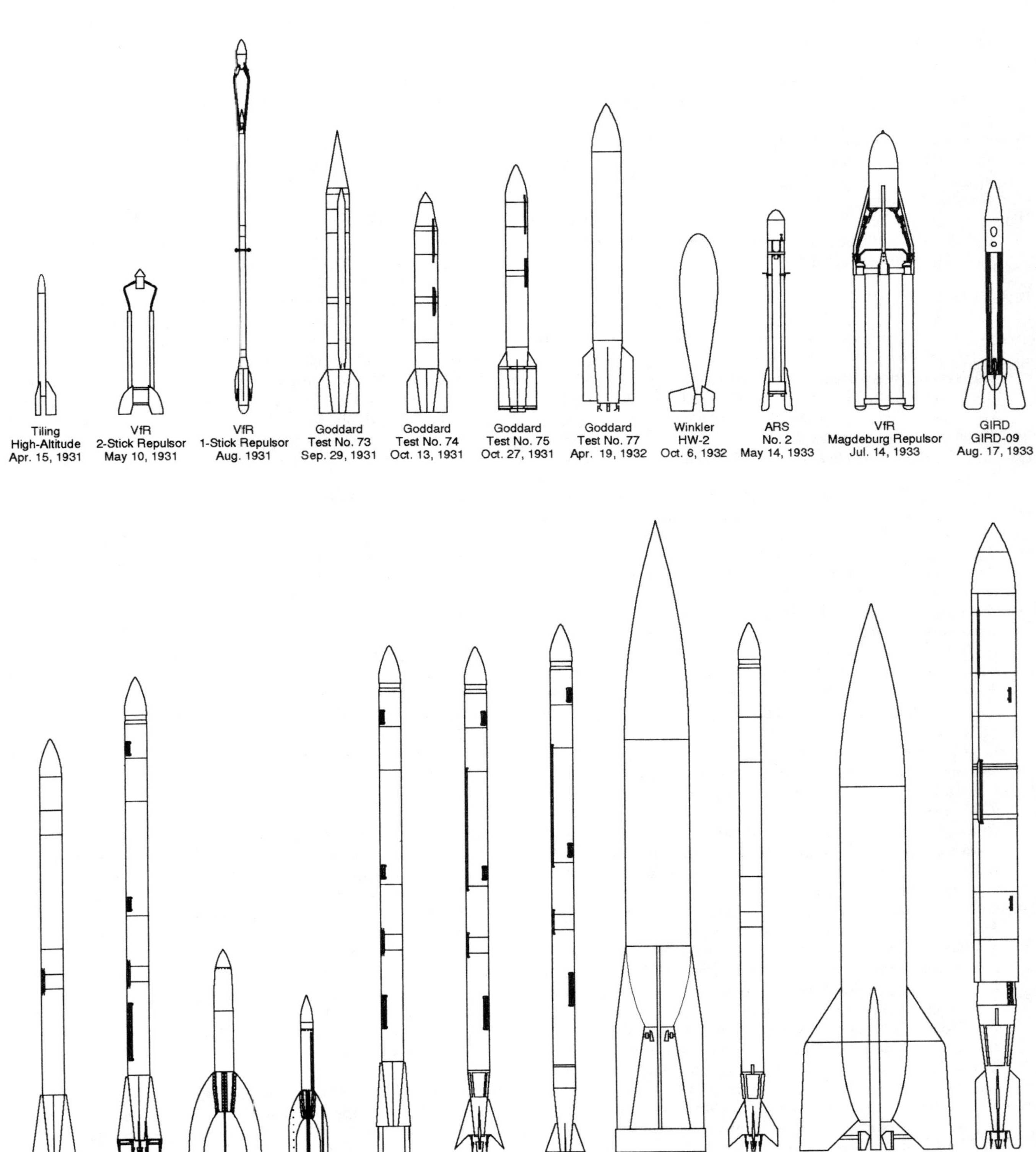

Bibliography

General Sources

Baker, David, *The Rocket,* Crown Publishers, 1978. An epic tome on the history of the rocket, this book covers rocket development in the US, Germany, and the Soviet Union.

Various authors, AAS History Series, American Astronautical Society. Beginning with Volume 6, these books contain the proceedings of International Academy of Astronautics history symposia beginning in 1967. The papers within are often unique sources of detailed information on obscure early rocket projects.

Winter, Frank H., *Prelude to the Space Age: The Rocket Societies: 1924-1940,* Smithsonian Institution Press, 1983. The stories of the ARS, VfR, GIRD, and others are all laid out in one place in this useful reference.

Wulforst, Harry, *The Rocketmakers,* Orion Books, 1990. A collection of vignettes in lives and works of leading American and German rocket pioneers.

Robert H. Goddard

Clark, Arthur C, and the editors of *Life, Man and Space,* Life Science Library, Time Life books, 1964, 1966. This book was widely distributed, and can be found in many attics and used bookstores. Its profile of Robert Goddard, illustrated with large-format photographs, introduced a generation of space enthusiasts to the work of Goddard.

Durant, F.C., *Robert H. Goddard: The Roswell Years (1930-1941),* Prepared for presentation at the Seventh International History of Astronautics Symposium of the International Academy of Astronautics, October 1973, Baku, USSR, Reproduction sponsored by Roswell Rotary Club and Roswell Museum and Art Center. This is a source of photographs and brief descriptions of many of Goddard's most important experiments. A version of this paper, with fewer photographs, appears in *History of Rocketry and Astronautics,* Kristan R. Lattu, ed., AAS History Series, Volume 8, American Astronautical Society/Univelt, 1989.

Goddard, Robert H., ed. by Esther Goddard and Edward Pendray, *Rocket Development,* 1948 (Prentice-Hall Spectrum edition, 1961). Extracted from Robert Goddard's notebooks, this book gives a good feel for the notebooks, and lays out a useful narrative of Goddard's later research. Summaries of each test, beginning in December of 1929, are given.

Goddard, Robert H., ed. by Esther Goddard and Edward Pendray, *The Papers of Robert H. Goddard.* This large, 3-volume set can be found in some larger libraries, and is an excellent reference on Goddard's life and works. Many of Goddard's most important publications are reprinted, and all known publications, including patents, are listed.

Goddard, Robert H., *A Method of Reaching Extreme Altitudes* and *Liquid Propellant Rocket Development.* These original Smithsonian monographs are rarities, but they were reprinted as a single volume, *Rockets,* which may be found at some larger libraries. The two are now available in the NASA publication *Exploring the Unknown: Selected Documents in the History of the US Civil Space Program, Volume I: Organizing for Exploration,* ed. by John M. Lodgeson, NASA SP-4407, 1995. Unfortunately, this edition omits photographs and suffers from misleading typesetting of equations.

Goddard, Robert H., *Development of Liquid Fueled Rocket,* and *Rocket Experiments.* These typed transcripts of Robert Goddard's notes are available to researchers by special appointment at the National Air and Space Museum library and the Goddard Library Special Collections. Drawings do not exist for most of Goddard's rockets, but the author has extracted dimensions from the text of these notebooks to produce most of the drawings in this book. The general public can get a taste of these documents by reading *Rocket Development.*

Lehman, Milton, *Robert H. Goddard: Pioneer of Space Research,* De Capo Press edition, 1988. (first published in 1963 as *This High Man*). This excellent biography will satisfy almost anyone interested in the life of Goddard. Well written and detailed, it is the definitive Goddard biography.

Goddard artifacts are displayed at several locations around the US. National Air and Space Museum, Washington, DC: 1918 smokeless powder rocket (1" recoiless gun projectile), in the Rocketry gallery; the rocket of May 4, 1926 (this model did not fly, but it includes components that flew on the previous 2 tests), in the Rocketry gallery; the Hoopskirt Rocket of December 26, 1928, in the Rocketry gallery; an A-series rocket (all components but the tanks have flown), at the Garber Facility; the tail section of a late L-series rocket (L-21 to L-30), in the Rocketry gallery; and rocket P-36, in the Milestones of Flight gallery. The Goddard Wing of the Roswell Museum: original launch tower as re-built for flight L-30 and the P-series tests, various rocket components. The Goddard exhibit room in the Goddard Library at Clark University: various components of Goddard rockets, as well as a model airplane with rocket-powered propellor blades.

The American Rocket Society

Pendray, Edward, *The Coming Age of Rocket Power*—Revised Edition, Harper & Brothers, 1947, includes summaries of the events recalled by one of the prime movers of the ARS.

Alfred Maul

Winter, Frank H., "Camera Rockets and Space Photography Concepts before World War II" *History of Rocketry and Astronautics,* Kristan R. Lattu, ed., AAS History Series, Volume 8, American Astronautical Society/Univelt, 1989, pp. 73-89. An excellent summary of early rocket-based photography, with photos of and from Maul's rockets.

The VfR

Gardner, Martin, *Fads and Fallacies in the Name of Science,* Dover Publications, 1957, pp. 22-27. Explains the genesis of the hollow-Earth theory that led to the Magdeburg rocket.

Ley, Willy, *Rockets, Missiles, and Space Travel,* Revised Edition, Viking Press, 1957. First-hand accounts of the birth, life, and death of the VfR. Also the prime source of information on Reinhold Tiling.

Sänger-Bredt, Irene, and Engel, Rolf, "The Development of Regeneratively Cooled Liquid Rocket Engines in Austria and Germany, 1926-1942," *First Steps Toward Space,* Frederick C. Durant III, George S. James, Editors, AAS History Series, Volume 6, American Astronautical Society/Univelt, 1985. pp. 217-246.

Von Braun, Wernher, and Ordway III, Frederick I., *History of Rocketry and Space Travel,* Third Revised Edition, Thomas Y. Crowell, 1975. Complements Ley's book, and also gives good coverage to American and Soviet work.

Von Braun and the German Army

Hölsken, Dieter, *V Missiles of the Third Reich,* Monogram Aviation Publications, 1994. Includes rare photographs of A-3 and A-5 flights, and cutaway views of the A-2, A-3, and A-5. Unfortunately, the specifications accompanying the cutaway drawings are scrambled.

Neufeld, Michael, *The Rocket and the Reich: Peenemünde and the Coming of the Ballistic Missile Era,* Harvard University Press, 1995. This book begins with a detailed history of the development of the V-2, providing detailed descriptions of the A-1, A-2, A-3, and A-5 programs.

Ordway III, Frederick I. and Sharpe, Mitchell, *The Rocket Team,* Thomas Y. Crowell, 1979. Rocket development in Germany and the US by Wernher von Braun's group.

GIRD

Biryukov, Yu. V., "The Role of Mikhail K. Tikhonravov in the development of Soviet Rocket and Space Technology," *History of Rocketry and Astronautics,* Kristan R. Lattu, ed., AAS History Series, Volume 8, American Astronautical Society/Univelt, 1989, pp. 343-349.

Dushkin, Leonid S., "Experimental Research and Design Planning in the Field of Liquid-Propellant Rocket Engines Conducted Between 1934-1944 by the Followers of F. A. Tsander," *History of Rocketry and Astronautics,* R. Cargill Hall, editor, AAS History Series, Volume 7, Part II, American Astronautical Society/Univelt, 1986, pp. 79-97.

Dushkin, Leonid S., and Moshkin, Yevgeny K., "Analysis of Liquid-Propellant Rocket Engines Designed by F. A. Tsander," *History of Rocketry and Astronautics,* AAS History Series, Volume 7, Part II, American Astronautical Society/Univelt, pp. 99-105.

Golanov, Yaroslav, *Sergei Korolev: The Apprenticeship of a Space Pioneer,* MIR Publishers, Moscow, 1975.

Stoiko, Michael, *Soviet Rocketry: Past, Present, and Future,* Holt, Rinehart, and Winston, 1970. Includes a rare description of the work of LenGIRD.

Merkulov, I. A., "Organization and Results of the Work of the First Scientific Centers for Rocket Technology in the USSR," *History of Rocketry and Astronautics,* Frederick I. Ordway, III., AAS History Series, Volume 9, American Astronautical Society/Univelt, 1989, pp. 63-77.

Oberg, James, *Red Star in Orbit,* Random House, 1981.

Rauchenbakh, Boris V., "Sergei Pavlovich Korolev (1907 to 1966)," *The Cambridge Encyclopedia of Space,* Michael Roycroft, editor, Cambridge University Press, 1990.

Riabchikov, Evgeny, *Russians in Space,* Novosti Press Agency, Moscow, and Doubleday & Co., 1971. Includes story of GIRD-09, and several photographs of GIRD, RNII, and KB-7 rockets.

Romanov, A., *Spacecraft Designer,* Novosti Press Agency Publishing House, Moscow, 1976. Includes anectdotes from the Korolev's GIRD period.

Rozkov, V, *Sportivnnye Modeli Raket,* Moscow, 1984. A Russian-language guide to model rocketry, it includes prototype and model drawings for several early Soviet rockets.

Tikhonravov, Mikhail K., "From the History of Early Soviet Liquid-Propellant Rockets," *First Steps Toward Space,* AAS History Series, Volume 6, American Astronautical Society/Univelt, pp. 287-293.

Tikhonravov, Mikhail K., and Zaytsev, V. P., "On the History of the Stratospheric Rocket Sonde in the USSR, 1933-1946," *History of Rocketry and Astronautics,* R. Cargill, Hall, ed., AAS History Series, Vol. 7, Part II, American Astronautical Society/Univelt, 1986, pp. 65-78.

RNII

Tikhonravov and Zaytsev, "On the History of..." pp. 70-72.

Leonid S. Dushkin, "Experimental Research and Design Planning in the Field of Liquid-Propellant Rocket Engines Conducted between 1934-1944 by the Followers of F. A. Tsander," *History of Rocketry and Astronautics,* R. Cargill, Hall, ed., AAS History Series, Vol. 7, Part II, American Astronautical Society/Univelt, 1986, pp. 81-83.

Polyarny, Korneyev, and KB-7

Merkulov, "Organization and Results..."

Polyarny, A. I., "On Some Work Done in Rocket Techniques, 1932-38," *First Steps Toward Space,* Frederick C. Durant III, George S. James, editors, AAS History Series, Volume 6, American Astronautical Society/Univelt, 1985, pp. 185-201

Model Rocketry

Alway, Peter, *The Art of Scale Model Rocketry,* Saturn Press, 1994. A guide specifically for designing, building and flying scale model rockets from prototype data like that provided in this volume. Includes plans for a smaller scale GIRD-09.

Stine, G. Harry, *Handbook of Model Rocketry,* Sixth Edition, Wiley, 1994. The most complete guide available to the hobby of flying model rocketry.

Van Milligan, Tim, *Model Rocket Design and Construction,* Kalmbach, 1995. This book by a former Estes Industries kit designer details principles and possibilities for flying model rocket design.

"The Empty Frame, obtained after a long time." The launcher of the first liquid-fueled rocket, shortly after the flight of March 16, 1926. (courtesy of Clark University Goddard Library Archives)

Index

A. I. Polyarny shows off his R-06 Osoaviakhim Rocket to members of the Stratospheric Committee. (Novosti photo)

Publications from Saturn Press

Rockets of the World (2nd edition).....$35.00

A book about space boosters and sounding rockets compiled with the modeler in mind. 200 versions of 137 rockets from 14 countries. Entries on each rocket includes historical summaries and dimensioned and color-keyed drawings. Over 180 B&W photos. 384 page hardcover.

The Art of Scale Model Rocketry....$15.00

A book about building flying scale models. Includes 13 model plans, tips on designing, building, and flying accurate, well-crafted models. Dozens of illustrations. 96 page wire-bound softcover lies flat for work and photocopying patterns.

Retro Rockets: Experimental Rockets 1926-1941....$18.00

A book about the early liquid-fueled rockets of Robert Goddard and his contemporaries in the US, Germany, and the Soviet Union. More than 30 rockets depicted with dimensioned, color-keyed drawings and photos for modelers. 96 page hardcover.

Rockets of the World Poster.....$10.00

A wall poster depicting 155 rockets at a constant 1/300 scale. Full color, sky blue background, 22" x 34". Rockets include those in the book, *Rockets of the World* and a couple dozen other vehicles.

Ordering information:

$10.00 minimum order. US orders add 10% shipping and handling (20% for priority mail) Michigan orders add an additional 6% Mich. sales tax. Foreign surface mail: 16% shipping and handling (surface mail) Add a $1.00 rolled mailing charge if you are ordering one or more posters. Please write for airmail rates. Pay by Check, Money Order, or Visa/MasterCard.

Saturn Press
PO Box 3709
Ann Arbor, MI 48106-3709
USA
(313) 677-2321

National Association of Rocketry Membership Application

Don't fly your rockets in a vacuum...
...join the NAR and fly with friends!

Membership in the National Association of Rocketry gives you access to all of these benefits:

- ***Sport Rocketry* Magazine & *The Model Rocketeer.*** These NAR publications contain rocket plans, NAR news, product reviews, manufacturer news, contest coverage, rocket photos, club events, construction techniques, and helpful tips. Submit your own photos and ideas to share with others!
- **NAR Sections** — Rocketry is much more fun when you do it with other rocketeers. NAR Sections are local clubs that hold launches, sponsor contests, run building sessions, publish newsletters, help beginners, maintain launch equipment, and obtain FAA waivers.
- **NAR Technical Services** — NARTS publishes rocket plans, scale data, technical reports, and education guides. They also sell rocket software, books, and NAR decals, pins, caps, and patches.
- **Hobby Liability Insurance** — NAR's optional liability insurance provides $1 Million of bodily injury/property damage protection in case of accident. This can be helpful in obtaining permission to use a flying field where an owner requires coverage.
- **Contests, Sport Launches, and Conventions** — Local, regional, and national events where rocket flyers meet and have fun. Compete against other rocketeers, show off your models, learn new skills, and meet new rocket friends. A blast!
- **High-Power** — Get certified to buy and fly up to 'L' motors!

The NAR is the official non-profit national organization for model rocketry. The NAR establishes safety codes, certifies rocket engines, sponsors educational programs and awards, certifies national records, works to reduce restrictive local, state, and federal regulations, and promotes all forms of safe rocketry. As a member, you help support all these important activities so that our hobby can grow and be enjoyed by all. Thank you!

NAR®

National Association of Rocketry 3/96
Membership Application

Date of Birth:____/____/____ Date:____________

Name:________________________

Address:________________________

City:____________ State:____ ZIP:______

Home Phone: (____)____________

AMOUNT ENCLOSED: $________________

Membership Category (check only one):

❑ Junior Membership (under 16 as of Jan. 1st) **$20.00**
❑ Leader Membership (under 21 as of Jan. 1st) **$20.00**
❑ Senior Membership (21 or over as of Jan. 1st) **$35.00**

Check Applicable Boxes:

❑ New Member ❑ Renewal: NAR#__________
❑ Teacher Packet ❑ Section #__________
❑ Hobby Liability Insurance (expires Dec. 31) **$21.00**
❑ Family Membership (One magazine per family . . . **–$12.00**
One family member must join at full price. Deduct $12.00 from member category for family membership.)

Rights, privileges, and responsibilities of membership begin upon acceptance of this application by the NAR. Prices and services subject to change without notice.

I pledge to conduct all my sport rocketry activities in compliance with the NAR Safety Codes.

✘________________________
Signature (required)

MAIL YOUR COMPLETED MEMBERSHIP FORM TO:
National Association of Rocketry, P.O. BOX 177, ALTOONA, WI 54720